TARGET ASSAYS for MODERN HERBICIDES and RELATED PHYTOTOXIC COMPOUNDS

TARGET ASSAYS for MODERN HERBICIDES and RELATED PHYTOTOXIC COMPOUNDS

Edited by

Peter Böger
Gerhard Sandmann

Lehrstuhl für Physiologie und Biochemie der Pflanzen
Universität Konstanz
Konstanz, Germany

LEWIS PUBLISHERS
Boca Raton Ann Arbor London Tokyo

Library of Congress Cataloging-in-Publication Data

Target assays for modern herbicides and related phytotoxic compounds /
edited by Peter Böger, Gerhard Sandmann.
p. cm.
Includes bibliographical references and indexes.
ISBN 0-87371-539-X
1. Plants, Effect of herbicides on. 2. Herbicides--Research.
3. Biological assay. I. Böger, Peter. II. Sandmann, Gerhard.
QK753.H45T37 1992
632′.954--dc20 92-26381
CIP

Direct all inquiries to CRC Press, Inc., 2000 Corporate Blvd., N.W., Boca Raton, Florida, 33431.

PRINTED IN THE UNITED STATES OF AMERICA
1 2 3 4 5 6 7 8 9 0

Printed on acid-free paper

Preface

This book presents assay systems to quantify the inhibitory activity of modern compounds which act upon specific targets in the plant cell. Using algal or higher-plant-model species as well as cell-free enzymatic systems, assays are described which can be performed with the modern equipment of the biochemical laboratory, without the need for special and sophisticated instruments.

The editors have been working on herbicide mode of action for a long time. Many colleagues, particularly from industry, visited our laboratory to practice some of the assays developed by us. From numerous contacts with experts in weed science and herbicide development, we are confident that the physiological and biochemical laboratory approach will gain importance when screening for activity of new compounds. Biochemical methods have increasing merits to evaluate analogues of known inhibitors and to obtain data on structure/activity relationships. The preparative chemist will be given a quick answer as to how his chemical modifications match with the assumed biological target.

Only assays for plant target sites for which herbicides and inhibitors are already known are described. Obviously, newly synthesized compounds or biogenic fermentation products should be checked first for established inhibition sites. Of course, many enzymatic reactions are conceivable which may be successfully inhibited by future herbicidal compounds. It is not intended to suggest new possible herbicide targets (as badly as these may be needed). By presenting updated methodology and procedures written by experts in the field, this book will assist colleagues and their technicians in the laboratory to run adequate experiments with established systems. The technical and experimental details we present here have not been published previously. The assays refer to pigment biosynthesis and degradation, photosynthetic electron transport, amino acid biosynthesis, fatty acid formation, and inhibition of cell division. The chapters are grouped in this order. Some model systems and miscellaneous topics are described. Some redundancy is deliberately included, e.g., on assays of acetyl-CoA carboxylase or protoporphyrinogen oxidase. A good method can still be improved and differences in the procedure will help to draw attention to its critical steps. Generally, a chapter should be complete without the reader needing to refer too often to other contributions.

Care has been taken to include contributions from laboratories in which the assay method in question has been developed or actively dealt with. The participation of younger colleagues still working at the ''bench'' will also ensure that the procedures outlined can be used with confidence. The editors instructed the authors to include — whenever possible — original data provided directly by the experiments. The reader will be able to find information he needs to perform and evaluate the assays on his own. He should feel free to write to the authors or the editors if trouble occurs with an assay described.

Colleagues from industry and universities have been asked to contribute to this book. Regrettably some did not have the time to describe their experiments. Nevertheless, most of the relevant laboratories around the world are represented. We believe that no assay topical and appropriate to obtain quantitative inhibition data on plant specific targets known for modern inhibitors is missing.

The editors gratefully acknowledge the efforts that the contributors have made toward sharing their experience. Thanks are also expressed to Lewis Publishers for offering this book at a reasonable price. We would like to see this book not only in the laboratories of established colleagues, but also in the hands of interested graduate students who are working on problems of chemical weed control.

Peter Böger
Gerhard Sandmann
Konstanz, Germany

Note on Nomenclature

Herbicides are given by their trivial and chemical names. Formulas can be found in *The Pesticide Manual,* 9th edition, C. R. Worthing and R. J. Hance, Eds., British Crop Protection Council, Farnham, Surrey (1991); *Herbicide Handbook of the Weed Science Society of America,* 6th edition, Weed Science Society of America, Champaign, Illinois (1989); or in *Short Review of Herbicides and PGRs,* S. Imamura, H. Taniyama, and T. Isogawa, Eds., Hodogaya Chemical Company Ltd., Tokyo, Japan (1991).

The Editors

Peter Böger trained at the Universities of Göttingen and Munich (biology and chemistry). He received a Ph.D. in 1963 at Göttingen and was a post-doctoral fellow at the C. F. Kettering Research Laboratory, Yellow Springs, Ohio, from 1964 to 1967, conducting research on the biochemistry of photosynthesis. From 1968 to 1972, Dr. Böger was Assistant Professor at the Universities of Cologne and Bochum, Germany, pursuing enzymological inhibitor studies in photosynthesis. In 1972, he became a Professor at the University of Konstanz and head of the Chair of Plant Physiology and Biochemistry in the Faculty of Biology.

During the last 15 years, Dr. Böger's main research interests have been focused on the mode of action of herbicides which interfere with chloroplast metabolism and pigment biosynthesis, as well as the regulation of nitrogen fixation in blue-green algae. He has authored more than 300 research papers and reviews.

Dr. Böger is a member of several scientific societies (the American Society of Plant Physiologists, the Weed Science Society of America, Deutsche Botanische Gesellschaft, and the Federation of European Societies of Plant Physiology). He is a member of the editorial boards of *Pesticide Biochemistry and Physiology, Physiologia Plantarum,* and *Zeitschrift fuer Naturforschung, Part C* and serves on evaluation committees of the Deutsche Forschungsgemeinschaft and governmental institutions. He lectures and visits at universities abroad, particularly in the U.S., Egypt, and Japan. Dr. Böger has organized several workshops on Photosynthetic Herbicides in Konstanz, 1979; Wageningen, 1983; Lake Placid, 1987; and Monheim, 1989. These contributions have been published in special issues of *Zeitschrift fuer Naturforschung*. He is also the editor of several books on herbicide biochemistry, including *Target Sites of Herbicide Action* (CRC Press, 1989).

Gerhard Sandmann trained at the Universities of Mainz and Frankfurt (biology and chemistry). He received a Ph.D. in 1978 at Frankfurt. Dr. Sandmann has been Assistant Professor in Konstanz at the Chair of Plant Physiology and Biochemistry. In 1982 he was a post-doctoral fellow with R. Malkin at the University of California at Berkeley (EPR measurements of photosynthetic redox components). In Konstanz, he developed methodology and cell-free systems on carotenogenesis and conducted research on the biochemistry of plant pigment formation. Dr. Sandmann has had more than 100 original contributions and overviews published. He is coeditor of *Target Sites of Herbicide Action* (CRC Press, 1989).

Contributors

Nikolaus Amrhein, Institut für Pflanzenwissenschaften, Eidgenössische Technische Hochschule, Sonneggstrasse 5, CH-8092 Zürich, Switzerland

Akihide Andoh, Naruto Research Center, Otsuka Chemical Company, Ltd., Satoura-cho, Naruto-City 772, Japan

Peter Babczinski, Bayer AG, Landwirtschaft-Umweltforschung, Pflanzenschutz-Zentrum Monheim, Bayerwerk, D-5090 Leverkusen, Germany

Jean-Pierre Blein, Laboratoire de Phytopharmacie, Institut National de la Recherche Agronomique, BV 1540, F-21034 Dijon, France

Peter Böger, Lehrstuhl für Physiologie und Biochemie der Pflanzen, Universität Konstanz, P.O. Box 5560, D-7750 Konstanz, Germany

Jean-Michel Camadro, Laboratoire de Biochimie des Porphyrines, Institut Jacques Monod, 2 Place Jussieu, F-75251 Paris 05, France

Reynold Chollet, Sandoz Agro AG, Agrobiological Research Station, CH-4108 Witterswil, Switzerland

Michel Couderchet, Lehrstuhl für Physiologie und Biochemie der Pflanzen, Universität Konstanz, D-7750 Konstanz, Germany

Stephen O. Duke, U.S. Department of Agriculture, Agricultural Research Service, P.O. Box 350, Stoneville, Mississippi 38776

Jörg Durner, Lehrstuhl für Physiologie und Biochemie der Pflanzen, Universität Konstanz, D-7750 Konstanz, Germany

Carl Fedtke, Geschäftsbereich Pflanzenschutz, Entwicklung/Herbizide, Pflanzenschutzzentrum Monheim, D-5090 Leverkusen-Bayerwerk, Germany

Manfred Focke, Botanisches Institut II, Universität Karlsruhe, Kaiserstrasse 12, D-7500 Karlsruhe, Germany

Valérie Gailus, Lehrstuhl für Physiologie und Biochemie der Pflanzen, Universität Konstanz, D-7750 Konstanz, Germany

Hermann M. Gleiter, Max-Volmer-Institut für Biophysikalische Chemie, Technische Universität Berlin, Strasse des 17. Juni 135, D-1000 Berlin 12, Germany

Simon P. Gough, Department of Physiology, Carlsberg Laboratory, Gamle Carlsberg Vej 10, DK-2500 Copenhagen Valby, Denmark

Klaus Grossmann, BASF Agricultural Research Station, D-6703 Limburgerhof, Germany

Gerhard Herrmann, Lehrstuhl für Physiologie und Biochemie der Pflanzen, Universität Konstanz, P.O. Box 5560, D-7750 Konstanz, Germany

F. Dan Hess, Research Division, Sandoz Agro, Inc., 975 California Avenue, Palo Alto, California 94304

Joy L. Howard, ICI Agrochemicals, Jealott's Hill Research Station, Bracknell, Berkshire, England

C. Gamini Kannangara, Department of Physiology, Carlsberg Laboratory, Gamle Carlsberg Vej 10, DK-2500 Copenhagen Valby, Denmark

William H. Kenyon, E.I. du Pont de Nemours and Company, Inc., Agricultural Products, Stine-Haskell Research Center, Newark, Delaware 19714

Takao Kida, Central Research Laboratories, Ajinomoto Company, Inc., 1-1 Suzuki-cho, Kawasaki-ku, Kanagawa 210, Japan

Oliver Knörzer, Lehrstuhl für Physiologie und Biochemie der Pflanzen, Universität Konstanz, D-7750 Konstanz, Germany

Pierre Labbe, Laboratoire de Biochimie des Porphyrines, Institut Jacques Monod, 2 Place Jussieu, F-75251 Paris 05, France

Hartmut K. Lichtenthaler, Botanisches Institut II, Universität Karlsruhe, Kaiserstrasse 12, D-7500 Karlsruhe, Germany

Remigius Manderscheid, Institut für Pflanzenökologie, Universität Giessen, Heinrich-Buff-Ring 38, D-6300 Giessen, Germany

Michel Matringe, Laboratoire de Physiologie Cellulaire Végétale, Université Joseph Fourier, CENG 85X, F-38041 Grenoble, France

Marie-Louise Milat, Laboratoire de Phytopharmacie, Institut National de la Recherche Agronomique, BV 1540, F-21034 Dijon, France

Beate Nicolaus, Lehrstuhl für Physiologie und Biochemie der Pflanzen, Universität Konstanz, P.O. Box 5560, D-7750 Konstanz, Germany

Gerhard Paul, BASF AG, Hauptlabor-A30, P.O. Box, D-6700 Ludwigshafen, Germany

Patrick Ravanel, Laboratoire de Physiologie Cellulaire Végétale, URA CNRS 576, Université Joseph Fourier, BP 53 X, F-38041 Grenoble, France

Gernot Renger, Max-Volmer-Institut für Biophysikalische Chemie, Technische Universität Berlin, Strasse des 17. Juni 135, D-1000 Berlin 12, Germany

Günter Retzlaff, BASF Agricultural Research Station, D-6703 Limburgerhof, Germany

Stuart M. Ridley, ICI Agrochemicals, Jealott's Hill Research Station, Bracknell, Berkshire, England

Thomas L. Rost, Section of Botany, University of California, Davis, California 95616-8537

Philippe Roy, Institut für Pflanzenwissenschaften, Eidgenössische Technische Hochschule, Sonneggstrasse 5, CH-8092 Zürich, Switzerland

Gerhard Sandmann, Lehrstuhl für Physiologie und Biochemie der Pflanzen, Universität Konstanz, P.O. Box 5560, D-7750 Konstanz, Germany

René Scalla, Laboratoire des Xénobiotiques, Institut National de la Recherche Agronomique, 180 Chemin de Tournefeuille, F-31931 Toulouse, France

Timothy D. Sherman, U.S. Department of Agriculture, Agricultural Research Service, Southern Weed Science Laboratory, P.O. Box 350, Stoneville, Mississippi 38776

Hiroshiro Shibai, Central Research Laboratories, Ajinomoto Company, Inc., 1-1 Suzuki-cho, Kawasaki-ku, Kanagawa 210, Japan

Andreas Thiel, Lehrstuhl für Physiologie und Biochemie der Pflanzen, Universität Konstanz, P.O. Box 5560, D-77 0 Konstanz, Germany

Michel Tissut, Laboratoire de Physiologie Cellulaire Végétale, URA CNRS 576, Université Joseph Fourier, BP 53X, F-38041 Grenoble, France

Kevin C. Vaughn, U.S. Department of Agriculture, Agricultural Research Service, Southern Weed Science Laboratory, P.O. Box 350, Stoneville, Mississippi 38776

Diter von Wettstein, Department of Physiology, Carlsberg Laboratory, Gamle Carlsberg Vej 10, DK-2500 Copenhagen Valby, Denmark

Ko Wakabayashi, Department of Agricultural Chemistry, Faculty of Agriculture, Tamagawa University, Machida-shi, Tokyo 194, Japan

Hiroyuki Watanabe, Agricultural Chemicals Laboratory, Research Center, Mitsubishi Kasei Corporation, 1000 Kamoshida-cho, Midori-ku, Yokohama, Kanagawa 227, Japan

Petra Weiss, Lehrstuhl für Physiologie und Biochemie der Pflanzen, Universtät Konstanz, D-7750 Konstanz, Germany

Daisuke Yanase, Naruto Research Center, Otsuka Chemical Company, Ltd., Satoura-cho, Naruto-City 772, Japan

Contents

PHOTOSYNTHETIC ELECTRON TRANSPORT

AMINO ACID BIOSYNTHESIS

FATTY ACID FORMATION

INHIBITION OF CELL DIVISION

MODEL SPECIES AND MISCELLANEOUS TOPICS

Pigment Biosynthesis and Degradation

CHAPTER 1

Spectral Determination of Carotenoid Precursors in *Scenedesmus* Cells Treated with Bleaching Herbicides

Gerhard Sandmann

INTRODUCTION

The so-called bleaching herbicides interfere with the biosynthesis of α- and β-carotene. Depending on the target enzyme of the herbicidal inhibitor, different carotene precursors of α- and β-carotene can be accumulated (see Sandmann and Böger, 1989, for review). As all these precursors differ by their spectral characteristics, simple optical absorbance spectra can reveal the inhibitory properties of a compound as well as the enzyme target. For these experiments, cells of *Scenedesmus acutus* are the most suitable because simple cell extracts are sufficient for spectral analysis of carotenoids, and even UV-absorbing carotenes can be detected without interference by other absorbing compounds. Furthermore, *Scenedesmus* gave reliable I_{50} values for inhibition of biosynthesis of yellow α- and β-carotenes as well as their corresponding xanthophylls. Determination of the I_{50} value of norflurazon is given as an example.

0-87371-539-X/93/$0.00 + $.50

PROCEDURE

Cell Growth

Scenedesmus acutus (strain 276-3a, Algal Culture Collection, University of Göttingen, D-3400 Göttingen, Germany) is grown in a medium consisting of KNO_3, 1.0 g/L; KH_2PO_4, 0.74 g/L; $Na_2HPO_4 \times 2\ H_2O$, 0.26 g/L; $MgSO_4 \times 7\ H_2O$, 50 mg/L; $CaCl_2$, 7.56 mg/L; $Fe(NO_3)_3 \times 9\ H_2O$, 14.5 mg/L; EDTA, 13.3 mg/L; $Al_2(SO_4)_3 \times 18\ H_2O$, 6.6 μg/L; $MnCl_2 \times 4\ H_2O$, 0.1 mg/L; $CoSO_4 \times 7\ H_2O$, 2.8 μg/L, $ZnSO_4 \times 7\ H_2O$, 6.3 μg/L; H_3BO_3, 30.6 μg/L; $(NH_4)_6Mo_7O_{24} \times 4\ H_2O$, 1.8 μg/L; NH_4VO_3, 2.9 μg/L; $NiSO_4 \times 6\ H_2O$, 26.3 μg/L; KJ, 0.25 mg/L; KBr, 0.24 mg/L; LiCl, 8.4 μg/L; $CuSO_4 \times 5\ H_2O$, 30 μg/L. The cultures were inoculated to an optical density at 950 nm of 0.5, which is equivalent to a packed cell volume (pcv) of about 0.5 μL/mL, and the herbicides were added from a stock solution prepared in methanol so that the final concentration of methanol was maintained below 1% in the medium. The 200-mL cultures are placed in a Kniese growth apparatus for 24 hr in light of 45 W/m^2 (corresponding to 10,000 lux), and a constant stream of air containing 5% CO_2 is bubbled through. The pcv of the cells in the suspension is determined in calibrated microcentrifuge tubes by centrifugation for 15 min in a swing-out rotor at full speed. Alternatively, the dry weight can be determined after filtering 1 mL of suspension through a cellulose acetate membrane filter with pores of 0.8 μm diameter and heating the filters with the cells overnight at 120°C in an oven.

Herbicides

The different bleaching herbicides used were norflurazon (4-chloro-5-methylamino-2-(3-trifluoromethylphenyl)-pyridazin-3(2H)one), SAN 380H (4-(2-chloro-2-propenoxy)-2-isopropylamino-6-methylpyrimidine), and CPTA (2-(4-chlorophenylthio)-triethylamine HCl). The chemical structures are shown by Böger and Sandmann (1990).

Equipment

For this assay a registrating spectrophotometer and a Kniese growth apparatus (see Chapter 8 for description) are required as the only necessary equipment.

Extraction of Carotenoids

All extraction steps should be carried out under dim light to avoid photodegradation of carotenoids. *Scenedesmus* suspension (10 mL) is centrifuged for 10 min at 10,000 × *g*, the pellet is resuspended in 20 mL of methanol, and 2 mL of 60% KOH (w/v) are added. Carotenoids are extracted at 65°C in darkness for 20 min. The extract is then poured into a separatory funnel which contains 25 mL of 10% diethyl ether in petrol, and 10 mL of saturated

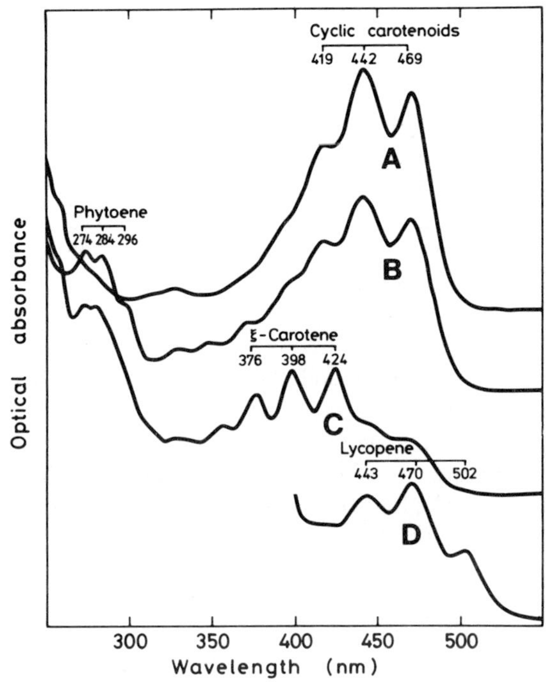

Figure 1. Absorbance spectra of *Scenedesmus* extracts from (A) untreated cultures and those treated with (B) 0.075 μ*M* norflurazon, (C) 0.3 μ*M* SAN 380H, or (D) 20 μ*M* CPTA.

NaCl solution is added and the upper organic layer is collected. Partitioning of the lower layer is repeated once. The volume of the combined organic phases is measured. Then the absorbance at 445 nm is read in a spectrophotometer and an absorbance spectrum is recorded.

Only when inhibitors are assayed in order to demonstrate their effect on lycopene accumulation must the following modification be applied: After methanol/KOH extraction of the cells, the suspension is centrifuged for 10 min at 10,000 × *g* and the resulted pellet re-extracted with acetone (20 min, 50°C). This suspension is then partitioned as described above, and a spectrum is recorded directly from the upper organic phase (see Figure 1D). This modification is avoidable if I_{50} values for inhibition of lycopene cyclase are determined.

RESULTS

Spectra of Cell Extracts

The original spectra of the pigment extracts from the control culture and from the ones treated with norflurazon (0.075 μ*M*), SAN 380H (0.3 μ*M*), and CPTA (20 μ*M*) are shown in Figure 1. The upper spectrum with the main

Table 1. Protocol for Determination of Colored Carotenoids in Norflurazon-Treated Cells

Norflurazon Concentration (*M*)	pcv[a] (μL/mL)	Extract Volume (mL)	E	Total Colored Carotenoids (μg/mL pcv)
0	3.8	39.5	0.495	2017
2.5×10^{-8}	3.4	41.5	0.285	1364
5.0×10^{-8}	4.7	41.5	0.301	1042
7.5×10^{-8}	4.8	42.0	0.236	810
10.0×10^{-8}	3.5	30.5	0.183	625

[a] Packed cell volume of the culture.

maximum at 442 nm in the control is an overlay of more or less identical spectra for α- and β-carotene and their xanthophylls. The corresponding absorbance is decreased in the presence of all three herbicidal compounds. With norflurazon a typical carotene-type spectrum, with main absorbance at 285 nm and two shoulders at 274 and 296 nm, can be found which is characteristic of phytoene, the first colorless carotene in the biosynthetic pathway. Treatment with SAN 380H also resulted in accumulation of some phytoene, but in addition absorbance at 376, 398, and 424 nm is detectable. This is the spectrum of ζ-carotene. In the acetone extracts of *Scenedesmus* cells treated with CPTA the spectrum of accumulated lycopene with peaks at 443, 470, and 502 nm can be recognized. It should be mentioned that in other algal species the UV-absorbing background may be too high to obtain a good phytoene spectrum.

Determination of the I_{50} Value for Norflurazon

Table 1 shows a typical protocol for the determination of colored carotenoids in norflurazon-treated cells. The concentration of total colored carotenoids of the cell suspension (in micrograms per milliliter) is quantitated by the equation

$$\text{Car} = 3.92\ E_{445} \times V/v$$

where V = volume of ether/petrol extract in milliliters
v = volume of sample in milliliters

The value is related to the pcv concentration of the culture. Alternatively, dry weight of cells can be used instead of pcv. Finally, reciprocal values of colored carotenoids per cells are plotted against herbicide concentration (Figure 2). In general, straight lines are obtained and the intercept with the abscissa gives the I_{50} concentration of the herbicide. The same procedure can be followed for other phytoene desaturase inhibitors and also for ζ-carotene desaturase inhibitors.

When I_{50} values are determined for inhibitors which accumulate lycopene (i.e., lycopene cyclase inhibitors) the following modifications must be made

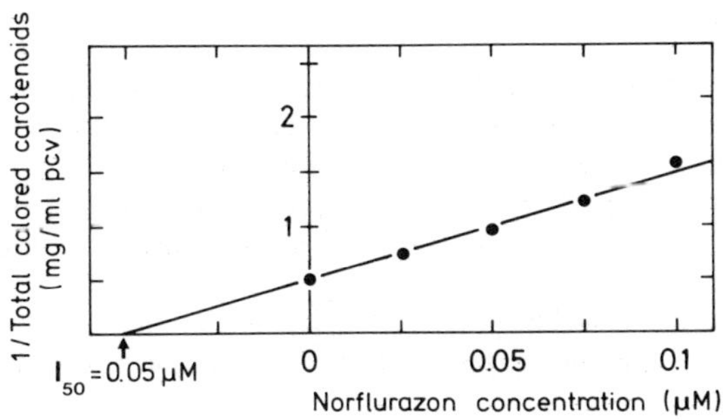

Figure 2. Determination of I_{50} value for formation of colored carotenoids in *Scenedesmus* treated with norflurazon.

due to the spectral contribution of lycopene at 445 nm. Absorbance is recorded at 445 and 505 nm; then the concentration of total cyclic colored carotenoids (in micrograms per milliliter) is calculated by the equation

$$Car = (7.25\ E_{445} - 3.31\ E_{505})$$

where V = volume of extract in milliliters
v = volume of sample in milliliters

COMMENTS

Presented is a simple procedure used to screen bleaching herbicides, revealing their targets in the biosynthetic pathway and enabling determination of the I_{50} values for specific mode of action. There is a fast equilibrium between the herbicide concentration outside and within the *Scenedesmus* cells, and no indications for degradation have been observed. Therefore, *Scenedesmus* is also suitable for assay systems to obtain quantitative data, as demonstrated. The I_{50} values proved to be very reliable, comparable to values obtained with higher plants (e.g., cress seedlings), and are very suitable for quantitative structure-activity relationship (QSAR) calculations. During the assay of numerous bleaching compounds in this laboratory, uptake problems for these derivatives have not encountered, with one exception, the bleaching triketones (Sandmann et al., 1990). This type of compound is inactive in single-cell systems of algae or higher-plant suspension cultures as well as in *in vitro* assays. Apparently, uptake and translocation by intact plants is a prerequisite for thier bleaching activity (see Chapter 2 on the cress seedling system for assay of benzoyl cyclohexanediones).

REFERENCES

Böger, P. and Sandmann, G., Modern herbicides affecting typical plant processes, in *Chemistry of Plant Protection,* Vol. 6, Bowers, W. S., Ebing, W., Martin, D., and Wegler, R., Eds., Springer-Verlag, Berlin, 1990, p. 173.

Sandmann, G. and Böger, P., Inhibition of carotenoid biosynthesis by herbicides, in *Target Sites of Herbicide Action,* Böger, P. and Sandmann, G., Eds., CRC Press, Boca Raton, FL, 1989, p. 25.

Sandmann, G., Böger, P. and Kumita, I., Atypical inhibition of phytoene desaturation by 2(-4-chloro-2-nitrobenzoyl)-5,5-dimethylcyclohexane-1,3-dione, *Pestic. Sci.,* 30, 353, 1990.

CHAPTER 2

Detection of Bleaching by Benzoyl Cyclohexanediones in Cress Seedlings by HPLC Separation and Determination of Carotenoids

Gerhard Sandmann

INTRODUCTION

Benzoyl cyclohexanediones (BCDs) belong to a special type of herbicide which affects the chloroplast pigments. They show bleaching effects and phytoene accumulation in higher plants when applied pre- or postemergent (Mayonado et al., 1989; Sandmann et al., 1990). The visible effect is similar to norflurazon. However, soybeans were only susceptible when germinated in the presence of this chemical class of herbicides. Soybean cell cultures or microalgae were unaffected by micromolar concentrations. Furthermore, *in vitro* phytoene desaturation with membranes obtained from chloroplasts, chromoplasts, and a cyanobacterium was not inhibited by a BCD (Sandmann et al., 1990).

For the assay of substituted BCDs a procedure is described which involves germination of cress seeds in the presence of herbicide and carotenoid analysis by high-pressure liquid chromatography (HPLC) of 3-day-old seedlings.

0-87371-539-X/93/$0.00 + $.50

PROCEDURE

Seedling Growth and Herbicide Treatment

Cress seeds (*Lepidium sativum,* about 100) are germinated in petri dishes on filter paper discs of 7 cm diameter in 6 mL of water to which the herbicide has been added from a methanol stock solution. The cultures are grown in complete darkness to avoid photo-oxidation of carotenoids by light. After 3 days, leaves are cut off and freeze-dried overnight.

Equipment

For the described assay a registrating spectrophotometer is needed for the recording of spectra. The HPLC system used is an isocratic one with a single pump connected to a programmable detector or one which can record at two wavelengths simultaneously (285 and 445 nm). Alternatively, two runs have to be carried out and recorded, one at 445 and the other at 285 nm. Finally, the peak area is determined by an integrator connected to the detector.

Herbicide

The BCD herbicide used throughout this study is the bleaching triketone 2-(4-chloro-2-nitrobenzoyl)-5,5-dimethylcyclohexane-1,3-dione (Sandmann et al., 1990).

Carotenoid Analysis

All work with carotenoid solutions has to be carried out in dim light. For carotenoid extraction 20 mL of methanol containing 6% KOH is added to about 10 to 50 mg ground leaf material and heated at 60°C for 20 min. The methanol extract is poured into a separatory funnel which contains 10 mL of 10% diethyl ether in petrol (bp 35 to 80°C), and then 10 mL of a saturated NaCl solution is added. The upper layer is collected and the lower one re-extracted with another 10 mL of diethyl ether/petrol. Total carotenoid content is determined in the combined upper layers by determination of the volume and measurement of the optical absorbance at 445 and 550 nm. The total amount of extracted colored carotenoids is calculated by the following equation:

$$\text{Car}\ (\mu\text{g}) = 3.92(E_{445} - E_{550}) \times V$$

where V = volume of diethyl ether/petrol extract in milliliters

After this determination the carotenoid solution is evaporated to dryness and resuspended in 200 μL of acetone. Carotenoids are separated with an isocratic HPLC system. The column used is a 25-cm Spherisorb ODS-1 5-μm

Table 1. Carotenoid Composition of Cress Leaves Treated with 2-(4-Chloro-2-Nitrobenzoyl)-5,5-Dimethylcyclohexane-1,3-Dione (BCD) as Determined by HPLC

Carotenoid (μg/g dry weight)	BCD Concentration (*M*)				
	0	10^{-7}	10^{-6}	5×10^{-6}	10^{-5}
Neoxanthin	4	2	2	<1	1
Violaxanthin	91	68	40	22	16
Antheraxanthin	11	12	6	4	3
Lutein	191	132	73	50	39
α-Carotene	2	5	2	1	<1
β-Carotene	33	22	8	3	2
Phytoene	0	36	82	140	148

column eluted with acetonitrile/methanol/2-propanol (85:10:5, v/v/v) at a flow of 1 mL/min (Ernst and Sandmann, 1988). For quantitation of carotenoids the integrator values of the areas were used. As the concentration of total yellow carotenoids of an injected sample was already determined, the relative integrator values could be taken to calculate the concentrations of individual yellow carotenoids, and quantitative calibration with standards was avoided. In the case of phytoene, the elution peak is collected and an optical absorbance spectrum is recorded. Absorbance maxima at 276, 285, and 298 nm demonstrate the identity of this colorless carotene. In addition, the absorbance value at 285 nm together with the volume of the eluted sample allows for the calculation of the amount of phytoene present in the sample used for HPLC separation.

The total amount of phytoene extracted from the cress leaves is calculated by the equation

$$\text{Phytoene } (\mu g) = \frac{E_{285} \times v_p}{75.7} \frac{V_E}{V_{inj}}$$

where v_p = the volume of the elution peak in microliters
V_E = the volume of the carotenoid extract subjected to HPLC (= 200 μL in the example described above)
V_{inj} = the volume injected (20 μL in our case)

By this quantitation, a calibration of the phytoene integrator values can be achieved.

If desired, the other carotenoids can be quantitated similarly. The amount of each of the yellow carotenoids listed in Table 1 can be calculated from one collected peak (e.g., lutein) by the equation

$$\text{Carotenoid } (\mu g) = \frac{E_{445} \times v_p}{250} \frac{V_E}{V_{inj}}$$

Table 2. Determination of Total Colored Carotenoids in Cress Leaves Treated with 2-(4-Chloro-2-Nitrobenzoyl)-5,5-Dimethylcyclohexane-1,3-Dione

	Dry Weight (mg)	Extraction Volume (mL)	E (445–550)	Total Colored Carotenoids (μg/mg dry weight)
Control	44.0	16.5	0.199	0.30
10^{-7} *M*	46.3	16.5	0.169	0.24
10^{-6} *M*	42.2	13.0	0.105	0.13
5×10^{-6} *M*	47.6	11.0	0.082	0.08
10^{-5} *M*	49.6	12.0	0.062	0.06

Note: E, optical density.

RESULTS

Table 2 shows a typical experiment for determination of total colored carotenoids in cress leaves treated with a BCD. Concentrations of up to 10^{-5} *M* BCD decrease the content of colored carotenoids almost to zero. From the calculated carotenoid concentrations an I_{50} value of 0.9 μ*M* can be estimated.

Figure 1 shows the separation of carotenoids in cress leaf extracts. The major carotenoid is lutein (peak 4, retention time, RT, = 8.6 min), followed by violaxanthin (peak 2, RT = 6.0 min), and β-carotene (peak 55, RT = 21.8 min). In addition, trace amounts of neoxanthin (peak 1, RT = 5.4 min), antheraxanthin (peak 3, RT = 7.2 min), and α-carotene (RT = 20.4 min) can be detected. In the presence of BCD all the peaks are decreased. Instead, a new compound, 15-*cis* phytoene (peak 6, RT = 26.9 min, absorbance maxima at 276, 285, and 298 nm), appears and is visible after changing the detection wavelength from 450 to 285 nm at 25 min. The quantitation of this determination of single carotenoids by HPLC is given is Table 1. With increasing concentrations of BCD, all yellow carotenoids decrease and their colorless precursor phytoene is accumulated gradually.

COMMENTS

As bleaching BCDs can only be assayed with higher plants, the procedure described offers a test system alternative to the previous one with algal cultures to obtain quantitative data on this type of herbicide. Accumulation of phytoene in cress seedlings grown in the dark gives strong evidence that BCDs are compounds which in principle, like norflurazon or related bleaching herbicides, inhibit phytoene desaturase. It should be pointed out that the I_{50} values are determined with dark-grown seedlings. Therefore, these values predominantly describe the inhibition of phytoene desaturase, and a contribution by additional photo-oxidative destruction of pigments is avoided. In general, I_{50} values determined with light-grown seedlings are lower. In addition, the phytoene values are also lower due to photodestruction during accumulation.

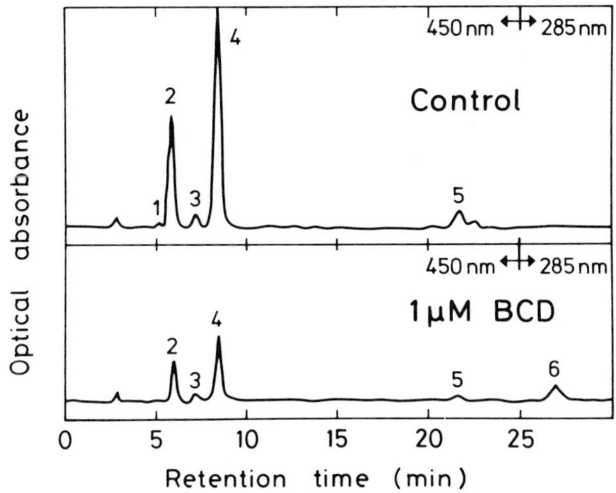

Figure 1. HPLC trace of a carotenoid extract from cress leaves treated without (control) or with 10^{-6} *M* substituted BCD; 1 = neoxanthin, 2 = violaxanthin, 3 = antheraxanthin, 4 = lutein, 5 = β-carotene, and 6 = phytoene.

REFERENCES

Ernst, S. and Sandmann, G., Poly-cis carotene pathway in the *Scenedesmus* mutant C-6D. *Arch. Microbiol.*, 150, 590, 1988.

Mayonado, D. J., Hatzios, K. K., Orcutt, D. M., and Wilson, H. P., Evaluation of the mechanism of action of the bleaching herbicide SC-0051 by HPLC analysis. *Pestic. Biochem. Physiol.*, 35, 138, 1989.

Sandmann, G., Böger, P., and Kumita, I., Atypical inhibition of phytoene desaturation by 2-(4-chloro-2-nitrobenzoyl)-5,5-dimethylcyclohexane-1,3-dione. *Pestic. Sci.*, 30, 353, 1990.

CHAPTER 3

In Vitro Assay System for Phytoene Desaturase Inhibitors with Isolated Thylakoids

Gerhard Sandmann

INTRODUCTION

The effect of bleaching herbicides on phytoene desaturation or on subsequent steps of the carotenoid biosynthetic pathway can be easily determined and quantitated by the two previously described assays in this book, which involve simple carotenoid analysis of treated algal cultures or plant seedlings. However, it may be of special interest to demonstrate that direct herbicide-enzyme interaction or *in vitro* inhibition values are required, e.g., in quantitative structure-activity investigations in which exclusively the structural requirements of herbicide binding to the enzyme is of interest (Sandmann et al., 1989; Sandmann et al., 1992). For this type of research an *in vitro* assay is needed. Furthermore, it has also been successfully employed to quantitate the contribution of phytoene desaturase inhibition to the overall herbicidal effect of a multifunctional herbicide (Sandmann and Böger, 1989). The *in vitro* reaction involves a coupling of two reaction sequences from two different organisms. The phytoene desaturase reaction is catalyzed by membranes from a cyanobacterium, and extracts from a fungal mutant blocked in geranylgeranyl pyrophosphate (GGPP) conversion provide ^{14}C-GGPP as substrate for the carotenogenic reactions from commercial ^{14}C-mevalonic acid.

0-87371-539-X/93/$0.00 + $.50

The assay described is the only one developed with photosynthetic membranes; in this case they are prepared from the unicellular cyanobacterium *Anacystis* PCC7942. Alternatively, the closely related *Aphanocapsa* PCC6714 can be employed. The two suitable fungal mutants are *Neurospora* al-2 (FGSC 313) from the Fungal Genetic Stock Center and *Fusarium* SG4 from the Departamento de Genética, Universidad de Sevilla. All organisms can be easily grown in 2 (cyanobacteria) to 4 (fungi) days. The *in vitro* reaction has proven extremely useful in characterizing numerous herbicidal compounds which interfere with carotenogenesis (Böger and Sandmann, 1990).

PROCEDURE

Growth of Organisms

Anacystis (Pasteur Culture Collection, no. 7942) is grown for 2 days at 32°C in a shaking incubator at a light intensity of 15 $\mu E/m^2 \cdot sec$. The BG11 medium consists of $NaNO_3$, 1.5 g/L; $K_2HPO_4 \times 3\ H_2O$, 52.5 mg/L; Na_2CO_3, 20 mg/L; $MgSO_4 \times 7\ H_2O$, 75 mg/L; $CaCl_2 \times 2\ H_2O$, 36 mg/L; ferric ammonium citrate, 6 mg/L; citric acid, 6 mg/L; EDTA, 0.1 mg/L; $Al_2(SO_4)_3 \times 18\ H_2O$, 6.6 μg/L; $MnCl_2 \times 4\ H_2O$, 0.1 mg/L; $CoSO_4 \times 7\ H_2O$, 2.8 μg/L, $ZnSO_4 \times 7\ H_2O$, 6.3 μg/L; H_3BO_3, 30.6 μg/L; $(NH_4)_6Mo_7O_{24} \times 4\ H_2O$, 1.8 μg/L; NH_4VO_3, 2.9 μg/L; $NiSO_4 \times 6\ H_2O$, 26.3 μg/L; KI, 0.25 mg/L; KBr, 0.24 mg/L; LiCl, 8.4 μg/L; $CuSO_4 \times 5\ H_2O$, 0.25 mg/L.

Both fungi can be grown in the same medium of 2.4% potato dextrose broth on a shaker. After 4 days the mycelium is harvested by filtration and lyophilized to absolute dryness. It then retains GGPP synthesizing activity for several weeks in the freezer.

Equipment

For the preparation of active *Anacystis* thylakoids a French pressure cell with the corresponding press is needed. The analysis of the reaction products is carried out with an isocratic HPLC system, connected to a programmable wavelength photometer and also on-line to a continuous liquid scintillation radioactivity monitor. As highly lipophilic compounds are separated, a liquid scintillation flow cell should be used instead of a solid scintillation cell to avoid background problems. Finally, all facilities necessary to work with ^{14}C-labeled compounds must be available.

Herbicides

The herbicide used as an example is norflurazon (= 4-chloro-5-methylamino-2-(3-trifluoromethylphenyl)-pyridazin-3(2*H*)one).

Preparation of Thylakoid Membranes and Fungal Extract

Thylakoid membranes are prepared from freshly grown *Anacystis* cultures by suspending the cell pellet from 0.5 L of culture (centrifugation at 5000 × *g* for 10 min) in 10 mL of 0.1 *M* Tris-HCl buffer, pH 8.0, containing 5 m*M* dithiothreitol. After French press treatment at 500 bar, the homogenate is centrifuged at 10,000 × *g* for 10 min and the pellet resuspended in 0.5 mL of the buffer mentioned above. For the experiments shown, cell extracts from *Fusarium* SG4 are obtained by rubbing 200 mg of freeze-dried mycelia through a sieve with a mesh size of 0.4 mm. Then 1 mL of 0.4 *M* Tris-HCl buffer, pH 8.0, containing 5 m*M* dithiothreitol is added, and the paste is stirred with a spatula for 1 min. After centrifugation (10,000 × *g*, 10 min), the supernatant is used in the coupled assay.

Incubation and Product Analysis

^{14}C-Mevalonic acid (MVA) was purchased from Amersham as the R-[2-^{14}C]mevalonic acid lactone with a specific radioactivity of about 50 mCi/mmol or 1.9 GBq/mmol (the R,S mixture is also suitable, but then twice the radioactivity must be added to the incubations because only the R isomer reacts with the enzyme systems). The lactone is supplied as a benzene solution (50 μCi per mL), and 5 μL are dried in a stream of N_2. Then the lactone is converted to the sodium salt of MVA by addition of 50 μL of 0.01 *N* NaOH. The assay medium contains the following, in a total volume of 0.5 mL: 5 μmol ATP, 1 μmol NAD, 3 μmol $MnCl_2$, 2 μmol $MgCl_2$ (all dissolved in water), 50 μL R-MVA solution (corresponding to 0.25 μCi or 5 nmol), 200 μL *Fusarium* SG4 membrane suspension (equivalent to 0.5 to 1.5 mg protein), and 200 μL *Anacystis* membrane suspension (equivalent to 100 μg chlorophyll) in the 0.1 *M* Tris-HCl buffer used above. The incubation is carried out for 2 hr in the light at 32°C. The reaction is terminated by adding 2.5 mL methanol and β-carotene (from Sigma Chemicals) and phytoene, 25 μg each, as marker carotenes (see Chapter 2 for isolation of phytoene). All subsequent steps are carried out in dim light to avoid degradation of carotenoid pigments. The content of the reaction tubes is poured into a separatory funnel with 15 mL 10% diethyl ether in petrol (bp 35 to 80°C), and after gentle shaking, 50 mL of a saturated NaCl solution is added. After partitioning, the upper diethyl ether/petrol layer is collected and evaporated in a stream of N_2. The dry residue is dissolved in 20 μL of acetone and injected for HPLC. Separation is carried out on a 25-cm Spherisorb ODS-1 5-μm column with acetonitrile/methanol/2-propanol (85:10:5, v/v/v) as eluent and a flow rate of 1 mL/min.

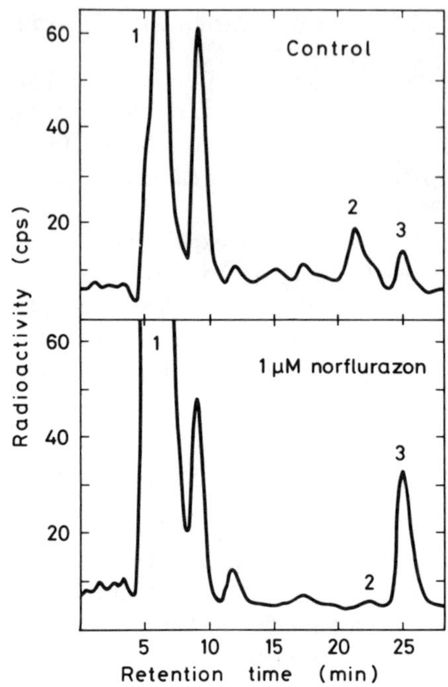

Figure 1. HPLC separation of ^{14}C-phytoene and ^{14}C-β-carotene from ^{14}C-GGPP by *Anacystis* thylakoids in the absence or presence of 1 μ*M* norflurazon. The radioactivity found in the control assay was 13,036 dpm for GGPP, 2,735 dpm for β-carotene, and 1,214 dpm for phytoene. In the norflurazon-containing assay, labeling was 15,204 dpm for GGPP, 186 dpm for β-carotene, and 3,518 dpm for phytoene. See text for details.

RESULTS

The HPLC trace of the radioactivity extracted after termination of the *in vitro* reaction shows four prominent peaks (Figure 1). The one at 5 min originates from the nonconverted substrate GGPP, which was hydrolized to its alcohol (GGOH). In the experiments presented here, *Fusarium* SG4 was used to generate ^{14}C-GGPP as substrate. The peaks at 21 and 25 min were identified as β-carotene (all-*trans* isomer) and phytoene (15-*cis* isomer), respectively, by co-chromatography with corresponding standards. The compound with retention times of 9 and 12 min are unknown but not related to the carotenogenic pathway. They also appear when the *in vitro* reaction is carried out with only the fungal extract. For a good *Fusarium* SG4 extract a yield of more than 10,000 dpm in GGPP is normal. When the *Anacystis* membranes show good activity, 2000 to 3000 dpm are incorporated into β-carotene. The radioactivity in β-carotene should be about twice that of phytoene. The β-carotene to phytoene ratio is a good indicator for the activity of phytoene desaturase in the preparation.

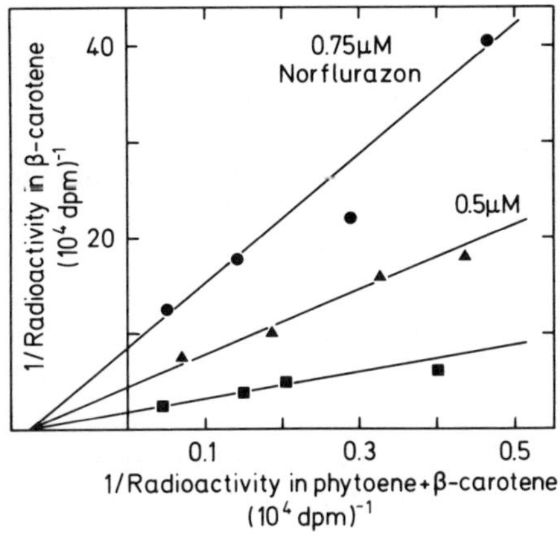

Figure 2. Double-reciprocal plot of substrate dependency of the phytoene desaturase reaction in the presence of norflurazon.

For almost complete inhibition of phytoene desaturase a concentration of 1 μM, which is about 10 times the k_i value (Sandmann et al., 1989), was employed. Under these conditions (Figure 1, lower trace), only trace amounts of β-carotene were formed by the cell-free system. The biosynthetic carotenogenic chain from GGPP is blocked at the level of phytoene, and the radioactivity accumulated in this pool is approximately the sum of the radioactivity found in phytoene and β-carotene in the control assay.

The *in vitro* system described is very useful for performing enzyme kinetic studies. As ^{14}C-phytoene, the substrate of the phytoene desaturase reaction, is generated in the course of the incubation, an average substrate concentration is calculated as the sum of radioactivity in phytoene + β-carotene. The contribution of the radioactivity in β-carotene can be regarded as the part of the substrate phytoene which has been converted in the biosynthetic chain. As shown in Figure 2, a Lineweaver-Burk plot of the calculated substrate values of phytoene desaturase and β-carotene, which is the metabolized product of the reaction, gives straight lines for all sets of experiments either with or without 0.5 or 0.75 μM norflurazon present. All three lines intersect at the same position of the abscissa. This result, together with reversibility of norflurazon binding (Sandmann et al., 1989), indicates that norflurazon is a noncompetitive inhibitor of phytoene desaturase. Consequently, its k_i value can be determined easily in a Dixon plot, which gives a value of 0.09 μM (Sandmann et al., 1989).

Examples for the use of the cell-free assay described here to correlate inhibition parameters with physiological properties of varied substituents of chemically different herbicides which interfere with phytoene desaturase are given by Sandmann et al. (1989 and 1991). The *in vitro* assay works well with all bleaching phytoene desaturase-inhibiting herbicides except for the benzoyl cyclohexanediones.

COMMENTS

It should be pointed out that the *in vitro* assay for phytoene desaturase described here is very complex and much more difficult to perform than the two previous assays (Chapters 1 and 2), in which only carotenoid analysis is involved. A prerequisite for its use is that experience in membrane isolation and biochemical routines in enzymological techniques are established in the laboratory. Growth of *Anacystis* and the fungal mutants is comparably simple. Both *Neurospora* al-2 (strain FGSC 313) and *Fusarium* SG4 catalyze the formation of ^{14}C-GGPP equally well. However, there is better access to the *Neurospora* mutant because it can be ordered from an official culture collection at the Fungal Genetic Stock Center by writing to Dr. Craig Wilson, Unviersity of Kansas, Department of Microbiology, 39th and Rainbow Boulevard, Kansas City, Kansas 66103.

REFERENCES

Böger, P. and Sandmann, G., Modern herbicides affecting typical plant processes, in *Chemistry of Plant Protection,* Vol. 6, Bowers, W. S., Ebing, W., Martin, D. and Wegler, R., Eds., Springer-Verlag, Berlin, 1990, p. 173.

Sandmann, G. and Böger, P., Inhibition of carotenoid biosynthesis by herbicides, in *Target Sites of Herbicide Action,* Böger, P. and Sandmann, G., Eds., CRC Press, Boca Raton, FL, 1989, p. 25.

Sandmann, G., Kowalczyk-Schröder, S., Taylor, H. M. and Böger, P., Quantitative structure-activity relationship of fluridone derivatives with phytoene desaturase. *Pestic. Biochem. Physiol.,* 42, 1, 1992.

Sandmann, G., Linden, H. and Böger, P., Enzyme-kinetic studies on the interaction of norflurazon with phytoene desaturase. *Z. Naturforsch.,* 44c, 787, 1989.

CHAPTER 4

Glutamate 1-Semialdehyde Aminotransferase as a Target for Herbicides

Simon P. Gough, C. Gamini Kannangara, and Diter von Wettstein

INTRODUCTION

The nonprotein amino acid δ-aminolevulinate is the common precursor for heme and chlorophyll biosynthesis. In chlorophyll-containing organisms and many bacteria, δ-aminolevulinate is synthesized from glutamate by the C_5 pathway. In contrast, animals and certain bacteria, e.g., *Rhizobium,* form δ-aminolevulinate from succinyl-CoA and glycine (Kannangara, 1991). The C_5 pathway of δ-aminolevulinate synthesis therefore constitutes a suitable target for the action of herbicides, which are nontoxic to animals and man. In these organisms, δ-aminolevulinate is formed in three steps. The first step is glutamyl-tRNA synthesis requiring chloroplast tRNA and its partner ligase. Because of its ubiquitous role in protein synthesis in living organisms the ligase is not a target for herbicide action. The following reactions are, however, specific to plants. Glutamate-tRNA is reduced to glutamate semialdehyde, which is transaminated to δ-aminolevulinate. The enzyme catalyzing this last step, glutamate semialdehyde aminotransferase, is the best studied of these two activities. It is thus a prime target for development of a broad-spectrum herbicide.

Glutamate 1-semialdehyde aminotransferase is an abundant protein in the stroma of greening plastids. It is encoded by a nuclear gene and has a molecular

0-87371-539-X/93/$0.00 + $.50

weight of 46 kDa. The holoenzyme consists of two identical subunits, and each subunit carries the cofactor pyridoxamine 5′-phosphate. Glutamate 1-semialdehyde aminotransferase has an exceptionally high affinity for L-glutamate 1-semialdehyde. The enzyme is inhibited by gabaculine (3-amino-2,3-dihydrobenzoic acid) and its analogue 6-amino-5-hydroxy-1,3-cyclohexadiene-1-carboxylic acid. In addition, acetylenic GABA (4-amino-hex-5-enoic acid), 4-amino-5-fluoropentanoic acid, and glutamic-γ-monohydroxamic acid are powerful inhibitors of this enzyme. Treatment of greening barley leaves with gabaculine and the other compounds reported here inhibits chlorophyll synthesis and causes the accumulation of the substrate glutamate 1-semialdehyde. In order to measure enzymatic activity *in vitro,* glutamate 1-semialdehyde has been chemically synthesized.

This chapter describes the preparation of the enzyme glutamate semialdehyde aminotransferase and its substrate and procedures for assaying the enzyme. Some examples of the action of presently known inhibitors of the enzyme are given.

PROCEDURE

A Protocol for Chemical Synthesis of Glutamate 1-Semialdehyde

Three methods are available for the synthesis of glutamate 1-semialdehyde (Kannangara and Gough, 1978; Houen et al., 1983; Gough et al., 1989). Two of these methods use N-carbobenzoxy-L-glutamic acid-γ-benzyl ester as starting material. Although these methods give the physiological isomer L-glutamate semialdehyde, they are difficult to perform and give low yields. The third method starts from 4-amino-hex-5-enoic acid. This compound is also called vinyl-GABA and is available as a mixture of D- and L-racemers from Merrill Dow, Strasbourg, France. An aqueous solution containing equimolar amounts of 4-amino hex-5-enoic acid and HCl is treated with ozone. This method directly forms the hydrochloride of glutamate 1-semialdehyde. It is analogous to an old method used for the formation of aminoacetaldehyde. The ozonized solution is then passed through a Dowex 50WX8 column in its H^+ form. This method gives pure D,L-glutamate 1-semialdehyde in yields of up to 80% (Figure 1). An ozone generator capable of producing 0.2 to 0.3 mmol of ozone per minute is required for this purpose. The rate of ozone production from the generator is determined after establishment of a constant flow in the following way. The ozone is bubbled through 100 mL of 10% KI for 1 min, and 5 μL of the solution is diluted to 3 mL with 10% KI and its optical density at 352 nm (A_{352}) is measured. The rate of ozone production in millimoles per minute is given by $2.29 \times A_{352}$.

A 10-mL test tube containing 1 mmol (129 mg) of vinyl-GABA dissolved in 1 mL of 1 *N* HCl is cooled in an ice bath. Ozone at the rate of 0.2 to 0.3 mmol/min is bubbled through this solution on ice for 8 to 10 min. Excess ozone tends to give lower yields and cause oxidation of the semialdehyde.

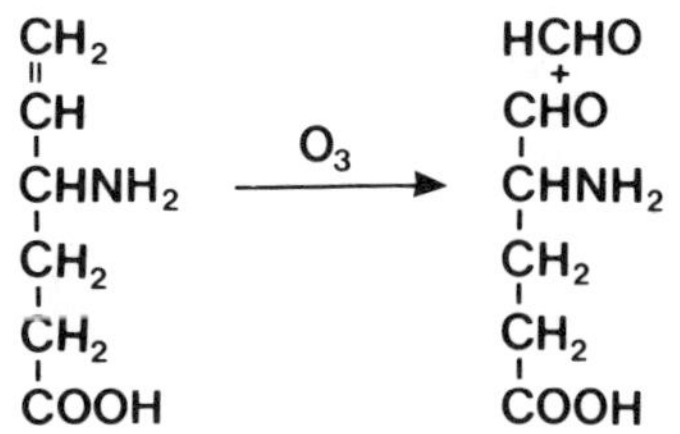

Figure 1. Chemical synthesis of glutamate 1-semialdehyde by reductive ozonolysis of vinyl-GABA.

The solution is immediately diluted with 20 mL of water and loaded onto a Dowex 50WX8 column pre-equilibrated with 0.05 *N* HCl. The cation exchange resin is pre-equilibrated by washing the resin with 1 *N* NaOH followed by 1 *N* HCl and then with 0.05 *N* HCl. The resin is packed in a BioRad Econocolumn to give a 1 × 8 cm bed. After the ozonized material is loaded, the column is washed with 20 mL of 0.05 *N* HCl. The glutamate 1-semialdehyde is then eluted with 0.5 *N* HCl and the eluate is collected in 2-mL fractions. Fractions (#5 to 16) containing glutamate 1-semialdehyde are pooled and freeze-dried in preweighed tubes, and the yield is determined by weighing. The dry solid D,L-glutamate 1-semialdehyde hydrochloride (molecular weight = 167) can be stored frozen at −20°C. The presence of glutamate 1-semialdehyde in the fractions can be determined prior to freeze-drying in the following way. Drops of each fraction are placed on filter paper, dried, and sprayed with ninhydrin. Aliquots (2 μL) from the ninhydrin-positive samples are chromatographed on silica gel 60 sheets with *n*-butanol:acetic acid:water (60:20:20 v:v:v) and sprayed with ninhydrin. Glutamate 1-semialdehyde migrates with an R_f of 0.54 (glutamic acid R_f = 0.22). The ninhydrin-positive fractions are assayed for substrate activity with a preparation of glutamate semialdehyde aminotransferase.

Ozone is a poisonous gas and will cause explosions if allowed to collect. Therefore, ozonolysis should be carried out in a fume cabinet. Ozone can be broken down by bubbling through 5% NaOH.

Preparation of Glutamate 1-Semialdehyde Aminotransferase

Extracts of chloroplast stroma prepared from plastids isolated from young and rapidly greening leaves contain active enzyme and are suitable for screening purposes. Leaves are homogenized in ice-cold 0.6 *M* glycerol containing 0.1 *M* Tricine-NaOH (pH 9.0) and filtered through a nylon gauze (31-μm mesh). The filtrate is centrifuged at 1500 × *g* for 5 min at 4°C. The pellet is enriched with plastids and is suspended in 0.3 *M* glycerol containing 0.1 *M* Tricine-NaOH (pH 7.9) and 1 m*M* dithiothreitol. The suspension is passed through a French pressure cell at 4°C and 8000 psi to break the plastids. The broken plastid preparation is centrifuged for 20 min in an ultracentrifuge at

$>100{,}000 \times g$. The clear supernatant contains the enzyme and can be used directly in the assays.

RESULTS

Enzyme Assay

Glutamate 1-semialdehyde aminotransferase is assayed by monitoring the production of δ-aminolevulinate from glutamate 1-semialdehyde in the presence and absence of enzyme.

The following solutions are prepared: 0.1 *M* phosphate buffer, pH 7.0 (or other suitable buffer), containing 1 m*M* dithiothreitol; 12 m*M* glutamate 1-semialdehyde HCl in water; 1 *M* levulinate adjusted to pH 7.0 with NaOH; and a protein extract with 2–4 mg protein per milliliter.

Mix separately in two tubes 1 mL of buffer, 10 μL of glutamate 1-semialdehyde HCl, and 25 μL of Na-levulinate. To one of the tubes add 50 to 200 μL of protein extract and to the other tube an equivalent volume of the buffer containing 0.3 *M* glycerol, 0.11 *M* Tricine-NaOH (pH 7.9) and 1.0 m*M* dithiothreitol. Incubate the tubes for 20 min at 30°C and stop the reactions by the addition of 50 μL of ice-cold 70% perchloric acid. Centrifuge at about $1000 \times g$ and transfer the clarified solutions to graduated tubes. Adjust the pH of these solutions to 6.9 using 0.5 *M* Na_3PO_4, add 100 μL of ethyl acetoacetate, and heat for 20 min in a boiling water bath. Cool the tubes to room temperature, adjust the volume to 1.5 mL with water, and add 1.5 mL of Ehrlich's reagent (1.82 g *p*-dimethylaminobenzaldehyde dissolved in a mixture containing 18 mL of 70% perchloric acid, 82 mL of glacial acetic acid, and 0.3 g of $HgCl_2$). After 20 min determine the absorption at 553 and 525 nm. Presence of active enzyme is indicated when the absorption obtained with enzyme is greater than that in the absence of enzyme and when the ratio of 525:553 is close to 0.66. The amount of δ-aminolevulinate is calculated using the following equation: nanomoles δ-aminolevulinate $= 41.7 \times (A_{553}$ with enzyme $- A_{553}$ without enzyme). More simplified assays are available for use with purified enzymes or with protein extracts exhibiting higher activity.

Inhibition of Glutamate 1-Semialdehyde Aminotransferase by Gabaculine, Glutamic-γ-Monohydroxamic Acid, and 6-Amino-5-Hydroxy-1,3-Cyclohexadiene-1-Carboxylic Acid

During enzyme catalysis the pyridoxamine 5′-phosphate in the aminotransferase is transiently converted to pyridoxal 5′-phosphate, which reacts with gabaculine to produce a stable compound, *m*-carboxyphenyl pyridoxamine 5′-phosphate. This adduct binds strongly to the enzyme and irreversibly inactivates it. Figure 2 illustrates the inhibition of glutamate 1-semialdehyde aminotransferase by three chemicals *in vitro*. The enzyme activity was in-

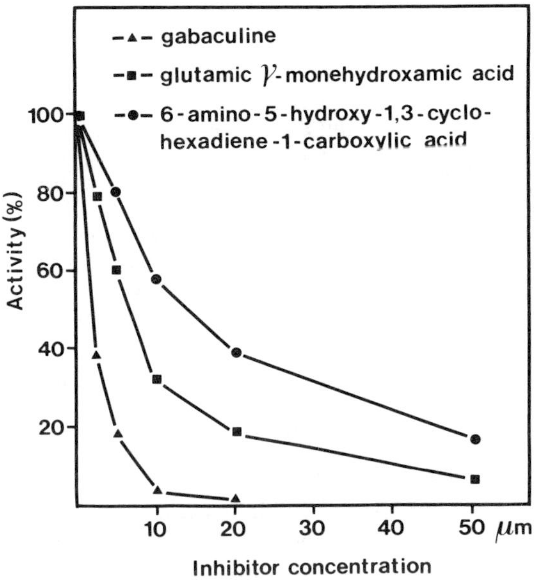

Figure 2. Inhibition of glutamate 1-semialdehyde aminotransferase *in vitro* by three potential herbicides. Gabaculine (3-amino-2,3-dihydrobenzoic acid) and glutamic-γ-monohydroxamic acid were purchased from Sigma Chemical Company. 6-Amino-5-hydroxy-1,3-cyclohexadiene-1-carboxylic acid was purified from culture filtrates of *Streptomyces aureofaciens* strain S652. (Modified from Kannagara, C. G. and Schouboe, A., *Carlsberg Res. Commun.*, 50, 179–191, 1985. With permission.)

hibited by 50% in the presence of 2 μ*M* gabaculine, 7.5 μ*M* glutamic-γ-monohydroxamic acid, of 17.5 μ*M* 6-amino-5-hydroxy-1,3-cyclohexadiene-1-carboxylic acid. These compounds also inhibit the enzyme *in vivo*. This is evident from the failure of etiolated leaves to turn green in light when they are treated with the compounds either by feeding from cut ends or by infiltration.

The procedure for feeding of etiolated barley leaves is as follows: The apical centimeter of the leaf is removed, and a clean cut is made at the base of the leaf to give an approximately 6-cm-long segment. The segments are placed upright in a beaker with 1 cm of their basal cut ends immersed in an aqueous solution of the inhibitor and illuminated. In the control experiments the basal ends of the leaf segments are immersed in water. A small fan is used to move the air above the leaves and facilitate uptake of solution. Figure 3 illustrates the inhibition of chlorophyll biosynthesis in etiolated leaves treated with 500 μ*M* gabaculine. When untreated etiolated leaf pieces are placed in light, rapid synthesis of chlorophyll occurs after an initial lag. Gabaculine feeding during illumination results in a severe inhibition of chlorophyll formation which is noticeable after 2 hr of illumination and becomes progres-

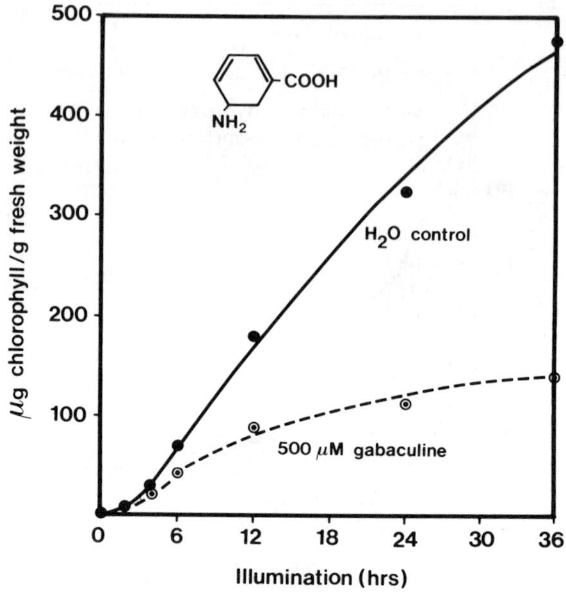

Figure 3. Inhibition of chlorophyll synthesis by gabaculine.

sively more pronounced with time. This procedure of using leaf segments is simple and can be used for testing other water-soluble compounds that are potential inhibitors of chlorophyll synthesis.

COMMENT ON THE USE OF GABACULINE AS AN HERBICIDE

Gabaculine is a neurotoxin in animals, including man. It is an inhibitor of γ-aminobutyrate transaminase, a pyridoxal enzyme (Rando, 1977). Gabaculine reacts irreversibly with the aldehyde group of pyridoxal-5-phosphate in this enzyme and irreversibly inhibits it. Therefore, gabaculine is not a practical herbicide. An analogue of gabaculine, however, or other specific inhibitors of glutamate 1-semialdehyde aminotransferase which are not toxic to mammals may prove useful.

REFERENCES

Gough, S. P., Kannangara, C. G. and Bock, K., A new method for the synthesis of glutamate 1-semialdehyde. Characterization of its structure in solution by NMR spectroscopy. *Carlsberg Res. Commun.*, 54, 99, 1989.

Houen, G., Gough, S. P., and Kannangara, C. G., δ-Aminolevulinate synthesis in greening barley. V. The structure and synthesis of glutamate 1-semialdehyde. *Carlsberg Res. Commun.*, 48, 567, 1983.

Kannangara, C. G. and Gough, S. P., Biosynthesis of δ-aminolevulinate in greening barley leaves. Glutamate 1-semialdehyde aminotransferase. *Carlsberg Res. Commun.,* 43, 185, 1978.

Kannangara, C. G., Biochemistry and molecular biology of chlorophyll synthesis, in *The Photosynthetic Apparatus: Molecular Biology and Operation,* Bogarad, L. and Vasil, I. K., Eds., Academic Press, San Diego, 1991, p. 302.

Rando, R. R., Mechanism of the irreversible inhibition of δ-aminobutyric acid: α-ketoglutaric acid transaminase by the neurotoxin gabaculine. *Biochemistry,* 16, 4604, 1977.

CHAPTER 5

Fluorometric Assay of Protoporphyrinogen Oxidase in Chloroplasts and in Plant, Yeast, and Mammalian Mitochondria

Jean-Michel Camadro, Michel Matringe, René Scalla, and Pierre Labbe

INTRODUCTION

The light-dependent phytotoxicity of diphenyl ethers (DPEs) and related herbicides is mediated by protoporphyrin IX, which accumulates in treated plants (Matringe and Scalla, 1987; Sandmann and Böger, 1988; Witkowski and Halling, 1988; Becerril and Duke, 1989) in response to the *in vivo* metabolic block in the heme and chlorophyll biosynthesis pathways. The block is at the level of protoporphyrinogen oxidase (EC 1.3.3.4), a membrane-bound enzyme which is strongly inhibited by DPEs (Matringe et al., 1989). Porphyrins are known to generate singlet oxygen in light; thus, their accumulation in treated plants explains the light-dependent peroxidation of membrane lipids and the ensuing disruption of membranes.

Protoporphyrinogen oxidase activity can be measured in both the mitochondrial and the chloroplast membrane fractions. This study describes a fluorimetric assay for protoporphyrinogen oxidase in mitochondrial and chloroplast membranes and the effects of DPEs on the enzyme *in vitro*. It is the first stage in an ongoing study of the structure-function relationships of available DPEs and a program to evaluate the herbicide potency of new DPE analogues.

0-87371-539-X/93/$0.00 + $.50

PROCEDURE

Preparation of Biological Samples

Corn seeds (*Zea mays* cv. Monclair) were surface sterilized with calcium hypochlorite, soaked in water for 1 day, and germinated on a stainless steel screen above distilled water for 6 to 7 days in darkness. Just before "etiochloroplast" isolation, the seedlings were allowed to green for 2 to 4 hr in dim light (30 μmol × m^{-2} sec^{-1} photosynthetically active radiation). Chloroplasts were isolated from etiolated leaves ground in a pH 7.7 buffer containing 0.5 *M* sucrose, 1 m*M* $MgCl_2$, 1 m*M* EDTA, 20 m*M* TES (2-(tris (hydroxymethyl)methylamino)-propane sulfonic acid), 10 mM HEPES (4-(2-hydroxyethyl)-1-piperazinepropane sulfonic acid), 5 m*M* cysteine, and 0.2% bovine serum albumin (BSA). The homogenate was centrifuged for 3 min at 1000 *g*; then the pellet was resuspended in the grinding buffer and centrifuged for 1 min at 150 × *g*. The supernatant was removed and centrifuged for 2 min at 2000 × *g*. The pellet was resuspended in grinding buffer containing 0.6 *M* sucrose, layered onto a cushion of 0.6 *M* sucrose buffer, and centrifuged for 15 min at 500 × *g*. That final pellet was used for protoporphyrinogen oxidase measurements.

Mitochondria were purified on Percoll gradients. Etiolated leaves were ground in a buffer containing 0.25 *M* sucrose, 10 m*M* MOPS (4-morpholinopropane sulfonic acid), 5 m*M* cysteine, and 0.2% BSA (pH 7.2). The resulting homogenate was centrifuged for 10 min at 2000 × *g*. The supernatant was recovered and centrifuged for 10 min at 10,000 × *g*. The pellet was gently Potter-homogenized in the grinding buffer and loaded on a discontinuous gradient of 15, 22.5, and 45% solutions of Percoll in grinding buffer. The tubes were centrifuged 30 min at 7500 × *g* and the mitochondrial fraction (at the interface of the 22.5 to 45% Percoll) was collected. This fraction was diluted in a mixture of 0.3 *M* mannitol, 10 m*M* EDTA, 20 m*M* MOPS, and 0.2% BSA (pH 7.2) and pelleted by centrifugation for 15 min at 11,000 × *g*.

Yeast (*Saccharomyces cerevisiae*) mitochondrial membranes were prepared from the haploid laboratory strain FL200 grown in a complete medium (yeast extract 1%, bactopeptone 1%, and glucose 2%). Commercial fresh baker's yeast was used, after three washings with water (5 min, 6000 × *g*) in some cases. A 60% (w/v) suspension of yeast cells in 0.1 *M* potassium phosphate buffer (pH 7.2) was disrupted by vigorous shaking for 5 min in 50-mL Corning tubes in a vortex in the presence of glass beads (0.45 to 0.5 mm diameter; 1 g/mL of suspension). The tubes were centrifuged for 5 min at 4000 × *g*, the pellets were washed twice with the homogenization buffer, the supernatants were pooled, and the mitochondrial membranes were pelleted by centrifugation for 30 min at 45,000 × *g* and resuspenced in the same buffer to a protein concentration of around 30 mg/mL.

Mouse (Swiss strain) liver mitochondria were isolated as follows: Mice were killed by cervical fracture, and the livers were sliced and washed to remove hemoglobin. They were homogenized in a Waring Blendor™ in a iso-osmotic buffer (0.35 *M* sucrose in 25 m*M* HEPES, pH 6.8). The cell-free fraction was centrifuged for 10 min at 3000 × *g* and the mitochondria were pelleted from the supernatant by centrifugation for 15 min at 25,000 × *g*. Mitochondrial membranes were washed twice in 50 m*M* potassium phosphate buffer (pH 7.2) by centrifugation (30 min at 100,000 × *g*) and resuspended in the same buffer to a protein concentration of around 30 mg/mL.

Equipment

Measurements were made in a Jobin-Yvon JY3D spectrofluorimeter equipped with a Hamamatsu R928H4 photomultiplier tube to enhance the sensitivity in the red region of the emission spectra of porphyrins. To avoid excess light on the cuvette leading to photo-oxidation of porphyrinogens and to ensure good selectivity, the band widths of the four slits were set as follows: 2 nm each for entry and exit for the excitation monochromator; 2 nm for entry and 20 nm for exit for the emission monochromator. The spectrofluorimeter was equipped with a thermostated cell holder and the cuvette contents were stirred magnetically. The kinetics of fluorescence emission and the emission and excitation spectra were recorded with an IFELEC 4000 recorder.

An Eppendorf 1101 M photometer equipped with the fluorimetric attachment and the recorder-adapter system was also found to be suitable. Identical results for protoporphyrinogen oxidase measurements were obtained using both fluorimeters by selecting one strong mercury emission line (406 nm) for excitation of porphyrins and by using relatively inexpensive optical filters from M.T.O. (11 rue Ampère, 91392 Massy Cedex, France; primary filter = Specivex A400b, secondary filter = Specivex DA633b).

Chemicals

The diphenyl ether-type herbicides used are AF(M), acifluorfen(-methyl) (methyl)-5-[2-chloro-4-(trifluoromethyl)-phenoxy]-2-nitrobenzoic acid; LS, LS 820340 {methyl-5-[2-chloro-4-(trifluoromethyl)-phenoxy]-2-chlorobenzoate}; RH, RH 5348 {methyl-5-[2-chloro-5-(trifluoromethyl)-phenoxy]-2-nitrobenzoic acid}.

Acifluorfen, sodium salt, was from ChemService, West Chester, Pennsylvania; AFM and LS 820340 were provided by Rhône Poulenc Agrochimie, France. RH 5348 was a gift from Rohm and Haas, Philadelphia, Pennsylvania. Protoporphyrin IX hydrochloride was from Porphyrin Products, Logan, Utah. Proporphyrin disodium salt was from Serva, Heidelberg, Germany.

Preparation of Protoporphyrinogen IX

All operations are carried out at room temperature under very dim light.

Protoporphyrinogen was prepared by chemical reduction of protoporphyrin with freshly prepared 3% sodium amalgam. Protoporphyrin hydrochloride (6 mg) was dissolved in 10 mL of a mixture of 0.04 *M* KOH in 20% ethanol to yield a stock solution (~1 m*M*) which was stored frozen as 0.4-mL aliquots in 1-mL glass tubes. Protoporphyrinogen was prepared by diluting one aliquot 2.5 fold (in a 10-mL Corning glass tube) with 0.2 *M* potassium phosphate buffer (pH 7.6) containing 50 m*M* L-ascorbic acid. This dark-brown solution was reduced with finely ground 3% sodium amalgam (2 g/mL) for 5 min with vigorous vortexing under a stream of nitrogen. The mixture was filtered through glass wool under nitrogen pressure and diluted to 4 mL (~0.1 m*M* protoporphyrinogen final concentration) with the phosphate-ascorbic acid buffer. In these conditions, protoporphyrin was reduced with high yields (>90%) without degradation. The fluorescence of the colorless (or slightly yellow) solution of protoporphyrinogen was negligible. Solutions of protoporphyrinogen were stable for several weeks if kept frozen in the dark as 0.4-mL aliquot fractions in 1-mL glass tubes under a drop of mineral oil. The exact concentration of protoporphyrinogen was determined spectrofluorometrically after total enzymatic oxidation to protoporphyrin by an excess of yeast mitochondrial membranes (prepared as described above) in the conditions of the enzymatic assay, by using the calibration curve (see below).

Assay of Protoporphyrinogen Oxidase Activity

Protoporphyrinogen IX oxidase activity was assayed spectrofluorometrically at 30°C by measuring the initial velocity of protoporphyrin IX formation from protoporphyrinogen IX (Labbe et al., 1985). Under our experimental conditions, the excitation and emission wavelengths were 410 and 633 nm, respectively. The standard reaction medium (1 mL) contained 0.1 mmol potassium phosphate reaction buffer saturated with air (final pH, 7.2), 1 μmol EDTA, 5 μmol dithiothreitol (DTT), 0.3 mg Tween 80, protoporphyrinogen, and 0.5 to 1 mg of membrane protein. Depending on the nature of the assay, the protoporphyrinogen IX concentration varied from 0.05 to 0.5 μ*M*. All inhibitors were dissolved in anhydrous dimethylsulfoxide, and the final solvent concentration was 1% (v/v) in all assays, including controls. The concentration of dissolved oxygen (the second substrate of the enzyme) was always saturating.

Standard Calibration Curve for Protoporphyrin Fluorescence Emission

A stock solution of protoporphyrin IX disodium salt (10 μ*M*) dissolved in 0.1 *M* Tris-HCl (pH 7.6) containing 1% (w/v) Tween 80 was used for flu-

Table 1. IC_{50} for AF, AFM, LS, and RH of Protoporphyrinogen Oxidase from Mouse Liver, Yeast, and Corn Mitochondria and Etiochloroplasts.[a]

Enzyme Origin	Mouse Mito.	Yeast Mito.	Corn Mito.	Corn Etio.
AF	250	9	500	500
AFM	19.5	7	6	4
LS	100	20	60	10
RH	3250	2000	700	180

[a] Mito., mitochondrial membranes; etio., etiochloroplastic membranes. For abbreviation of inhibitors see under chemicals.

orescence calibration experiments in the experimental conditions described above. The precise concentration of this stock solution was determined by measuring the absorbance at 409 nm ($\epsilon_m M = 262$) of a 10-fold dilution in 3 *M* HCl. The emission signal intensity increased linearly with protoporphyrin concentration up to 0.5 μ*M*. In this laboratory the curves have been extremely reproducible over a 10-year period.

RESULTS

Table 1 summarizes the IC_{50} values for AF, AFM, LS, and RH on the different sources of enzyme; IC_{50} is the inhibitor concentration producing a 50% inhibition under initial velocity conditions with saturating substrate concentration (Camadro et al., 1991).

COMMENTS

The main problem in protoporphyrinogen oxidase assays is the quality of protoporphyrinogen IX. The commercial origin of protoporphyrin (Porphyrin Products) and the method of reduction itself are crucial for protoporphyrinogen preparation and conservation. Careful training is required to obtain a **constant** satisfactory quality of protoporphyrinogen. Only freshly prepared sodium amalgam can be used, but ''environmental'' factors such as the relative atmosphere humidity may interfere significantly with the preparation of the amalgam.

Protoporphyrinogen oxidase activity can be measured very easily in the **mitochondrial membrane fractions** of plants, mammals, or microorganisms by measuring the initial velocity of the rate of protoporphyrin IX formation from protoporphyrinogen IX. However, since protoporphyrinogen oxidase also uses molecular oxygen as its second substrate, care must be taken to avoid anaerobiosis during the assay (1) by saturating the reaction buffer with air, (2) by extensive washings of the membrane fraction to remove endogenous respiratory substrates, and (3) by using small volumes (<100 μL) of biological sample in the 1-mL assays. The high concentration of EDTA used for the enzymatic assay (1 m*M*) is necessary to avoid any further metabolism of the protoporphyrin by membrane-bound metal chelatase(s).

The assay of protoporphyrinogen oxidase in etioplasts or chloroplasts is slightly more complex, since protoporphyrin IX fluorescence is quenched by chlorophylls or chlorophyll precursors present in the enzyme fractions. This problem can be overcome by quantifying the quenching of known concentrations of standard protoporphyrin IX by the enzyme fraction.

Despite these potential pitfalls, the fluorimetric assay of protoporphyrinogen oxidase is highly sensitive and reproducible.

REFERENCES

Becerril, J. M. and Duke, S. O., Protoporphyrin IX content correlates with activity of photobleaching herbicides. *Plant Physiol.,* 90, 1175, 1989.

Camadro, J. M., Matringe, M., Scalla, R., and Labbe, P., Kinetic studies on protoporphyrinogen oxidase inhibition by diphenyl ether herbicides. *Biochem. J.,* 277, 17, 1991.

Labbe, P., Camadro, J. M., and Chambon, H., Fluorometric assays for coproporphyrinogen oxidase and protoporphyrinogen oxidase. *Anal. Biochem.,* 149, 248, 1985.

Matringe, M. and Scalla, R., Induction of tetrapyrrole accumulation by diphenylether type herbicides, *Proc. Br. Crop. Prot. Conf. Weeds,* 3, 981, 1987.

Matringe, M., Camadro, J. M., Labbe, P., and Scalla, R., Protoporphyrinogen oxidase as a molecular target for diphenyl ether herbicides. *Biochem. J.,* 260, 231, 1989.

Sandman, G. and Böger, P., Accumulation of protoporphyrin IX in the presence of peroxidizing herbicides. *Z. Naturforsch.,* 43c, 699, 1988.

Witkowski, D. A. and Halling, B. P., Accumulation of photodynamic tetrapyrroles induced by acifluorfen-methyl. *Plant Physiol.,* 87, 632, 1988.

CHAPTER 6

Inhibition of Protoporphyrinogen Oxidase from Maize

Beate Nicolaus, Gerhard Sandmann, and Peter Böger

INTRODUCTION

The oxidation of protoporphyrinogen to protoporphyrin IX, the last common step in chlorophyll and heme biosynthesis, is catalyzed by the enzyme protoporphyrinogen oxidase (protoporphyrin IX:oxygen oxidoreductase, EC 1.3.3.4). This enzyme is the target of a large number of chemically different herbicides which cause degradative damage in plants and algae in the light (Matringe et al., 1989; Sandmann and Böger, 1988). The inhibition of protoporphyrinogen oxidase by these peroxidizers induces the accumulation of tetrapyrroles, mainly protoporphyrin IX, which can act as a photosensitizer (Becerril and Duke, 1989). The generation of singlet oxygen and radical intermediates in the light by the photoreactive protoporphyrin leads to the degradation of cell constituents like polyunsaturated membrane fatty acids, chlorophylls, and proteins.

The assay of protoporphyrinogen oxidase is based on the fluorescence of the reaction product protoporphyrin IX. Therefore, enzyme activity can be determined directly with a spectrofluorometer by monitoring the formation of protoporphyrin IX. Using this enzyme assay most of the peroxidizing compounds in plants were shown to be potent inhibitors of the protoporphyrinogen oxidase reaction in maize, with I_{50} values of 10^{-8} to 10^{-9} M. It

0-87371-539-X/93/$0.00 + $.50

could be shown that inhibition of this enzyme correlates quantitatively with ethane formation, photobleaching, and other herbicidal effects (see Chapters 7 and 8).

PROCEDURE

Plant Material

Maize seeds (*Zea mays* cv. Anjou) were soaked in water for 6 hr and germinated on vermiculite for 6 days in darkness at 30°C. Before harvesting the seedlings were allowed to green for 2 to 4 hr in the light (300 μEinstein (E)/m^2 × sec). Etioplasts were prepared by differential centrifugation steps to separate most of the glyoxysomes and mitochondria from etioplasts. After homogenizing 400 g of maize seedlings in 0.05 *M* tris-(hydroxymethyl)aminomethane (Tris-HCl), pH 7.3, containing 0.5 *M* sucrose, 1 m*M* $MgCl_2$, 1 m*M* EDTA, and 0.2% bovine serum albumin (w/v), the filtrate was centrifuged for 5 min at 10,000 × *g*. The resulting pellet was resuspended in about 25 mL of homogenizing buffer and centrifuged for 2 min at 150 × *g* to remove cell debris. Centrifugation of the supernatant for 15 min at 4000 × *g* resulted in 1 to 2 mL of purified plastids which could be stored in 20% (v/v) glycerol under liquid nitrogen for several weeks without significant loss of activity. Protein content of etioplasts was determined after the method of Lowry using bovine serum albumin as the standard.

Equipment

A Hitachi F2000 fluorescence spectrophotometer with a thermostated cell holder was used for the on-line determination of protoporphyrin IX formation. The excitation and emission wavelengths were set at 405 and 632 nm, respectively, and the fluorescence was recorded after starting the assay by addition of protoporphyrinogen. During the first 5 to 10 min both the protoporphyrinogen oxidase activity and the nonenzymatic conversion of protoporphyrinogen were found to be linear.

Herbicides

Oxyfluorfen, 2-chloro-1-(3-ethoxy-4-nitrophenoxy)-4-(trifluoromethyl)benzene, was supplied by Rohm and Haas, Philadelphia, Pennsylvania. Chlorophthalim, *N*-(4-chlorophenyl)-3,4,5,6-tetrahydrophthalimide, was a gift from Mitsubishi Kasei Corp., Yokohama, Japan. All compounds were dissolved in dimethylsulfoxide (DMSO).

Preparation of Protoporphyrinogen IX

Each step of the protoporphyrinogen preparation should be carried out in the dark. A 1 m*M* stock solution of protoporphyrin IX, free acid (Sigma), dissolved in 0.02 *N* KOH containing 20% ethanol was diluted to a 0.5 m*M* "working solution" with 0.1 *M* 2-(*N*-morpholino)ethane sulfonic acid (MES) buffer, pH 4.5, containing 50 m*M* ascorbic acid. The correct final concentration of protoporphyrin IX was determined by measurement of the absorbance of a 500-fold dilution of the working solution in 3 *N* HCl at 409 nm and calculated using the protoporphyrin extinction coefficient (= 262 mM^{-1} cm^{-1}). To prepare protoporphyrinogen, sodium amalgam was produced freshly under a stream of nitrogen. A total of 13 g of mercury and 0.5 g of thin slices of sodium were placed in a 500-mL round bottom flask equilibrated for 5 min with nitrogen. Generally, after 2 to 5 min of shaking the sodium with the mercury, a violet reaction occurred producing a thin layer of sodium amalgam which could be scratched off and ground to a fine powder. The protoporphyrin working solution was then reduced with the amalgam powder by vigorous shaking under a stream of nitrogen in the dark for 2 to 5 min until the dark red color disappeared, indicating the reduction of protoporphyrin. Up to 10 g amalgam per milliliter were used for the reaction. The protoporphyrinogen-containing solution was immediately filtered through several layers of glass wool and diluted two- to fourfold with MES buffer. The concentration of the substrate was determined by reoxidation of the protoporphyrinogen with 0.1% methyl ethyl ketone peroxide (dissolved in ethanol, v/v) and recording the amount of protoporphyrin obtained. Generally, 80 to 90% of the protoporphyrin IX was converted, resulting in a 0.1 to 0.2 m*M* protoporphyrinogen solution which could be stored in aliquots under a thin layer of paraffin in liquid nitrogen for several weeks.

Assay of Protoporphyrinogen Oxidase Activity

Protoporphyrinogen oxidase activity was measured at 30°C by following directly the protoporphyrin formation under initial velocity conditions. Standard assay conditions contained in a final reaction volume of 3 mL: 0.1 *M* Tris-HCl (pH 7.3), 1 m*M* EDTA, 5 m*M* dithiothreitol (DTT), 0.03% Tween 80 (w/v), 0.3 to 0.6 mg of etioplast protein, and 2 to 5 μ*M* protoporphyrinogen. The addition of the alkaline protoporphyrinogen to the assay mixture resulted in a final pH of 7.5, which has been found to be the pH optimum for maize protoporphyrinogen oxidase. The amount of protoporphyrin formed was calculated by using a protoporphyrin IX standard dissolved in 1% Tween 80 including 0.1 *M* Tris-HCl (pH 7.5). The rate of nonenzymatic oxidation of protoporphyrinogen, determined with a heat-denatured and sonicated protein, was subtracted from the enzyme activity.

RESULTS

Inhibition of Protoporphyrinogen Oxidase Activity

Figure 1 shows a typical measurement of enzyme activity without an inhibitor (control) and with increasing concentrations of oxyfluorfen from 0.5 n*M*

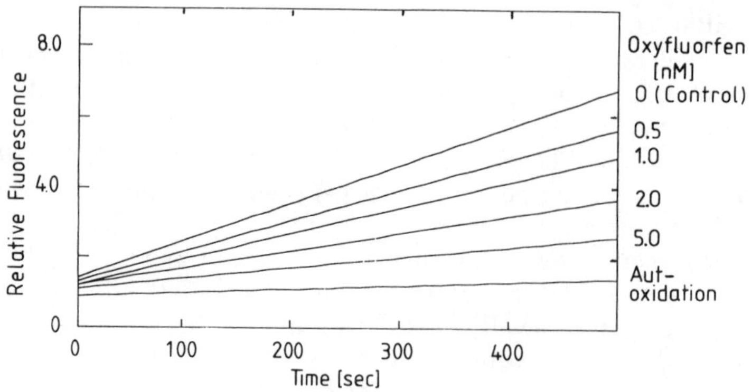

Figure 1. Direct measurement of protoporphyrin formation under initial velocity conditions from maize etioplasts (0.46 mg of protein per 3 mL assay volume) with various concentrations of the inhibitor oxyfluorfen.

up to 5 n*M*. The bottom line (= autoxidation rate) was obtained from a heat-denatured protein. The calculation of the activities in percent of control (control activity was 15.2 nmol/hr × mg protein) resulted in an inhibition pattern which is shown for two phytotoxic compounds in Figure 2A. Maximum inhibition of the enzyme was achieved at concentrations of 10^{-7} *M* for oxyfluorfen and 10^{-6} *M* for chlorophthalim. Oxyfluorfen showed a stronger inhibitory effect on protoporphyrinogen oxidase than chlorophthalim, which was also less active in inducing membrane degradation and bleaching of algal cells *in vivo*. The plot of inhibitor concentration vs. the reciprocal value of the rate of protoporphyrin formation (in percent of the untreated control) resulted in straight lines (Figure 2B). From the intersection with the abscissa, I_{50} values of 1.5 and 26 n*M* for oxyfluorfen and chlorophthalim, respectively, were obtained.

Determination of Inhibition Characteristics

The kinetic analysis of the inhibition of protoporphyrinogen oxidase by oxyfluorfen is shown in Figure 3. The initial velocity of protoporphyrin formation vs. the substrate concentration is given in the double-reciprocal Lineweaver-Burk plot. The lines obtained with various concentrations of oxyfluorfen intersect on the vertical axis, indicative of a competitive inhibition. The secondary plots of the slope of the lines vs. the inhibitor concentrations yield an inhibitor constant of 0.2 n*M* for oxyflurofen.

COMMENTS

A direct assay of the target enzyme of peroxidative compounds is described which makes it possible to gain insight into the molecular action of peroxidizing herbicides which induce radicalic degradation in plants. The measure-

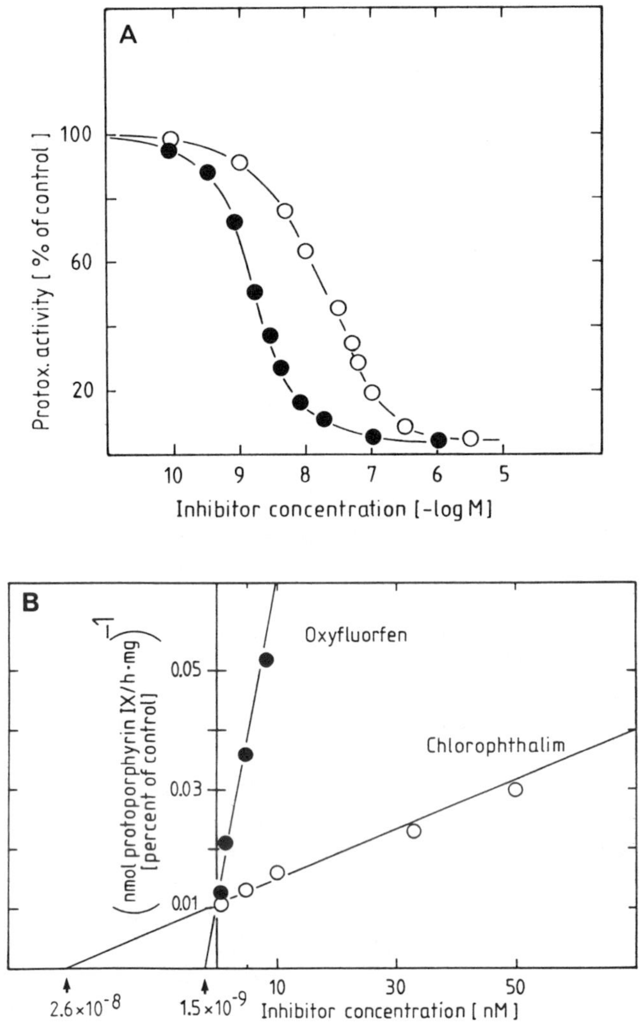

Figure 2. (A) Inhibition of protoporphyrinogen oxidase (protox) by oxyfluorfen (●-●) and chlorophthalim (○-○) in crude maize etioplasts, and (B) determination of I_{50} values by the reciprocal plot of protoporphyrin IX formation (in percent of control) vs. inhibitor concentration. Control specific activity was 15.2 nmol of protoporphyrin per hour per milligram protein.

ment of protoporphyrinogen oxidase activity *in vitro* is complicated by the presence of chlorophyll in the assay mixture. At high chlorophyll concentrations significant quenching of fluorescence occurs, limiting the amount of green plant thylakoids that can be used in the test. In the case of maize etioplasts, the highest specific enzyme activities were obtained in thylakoid fractions illuminated for 2 to 5 hr, which allowed detection of protoporphyrin

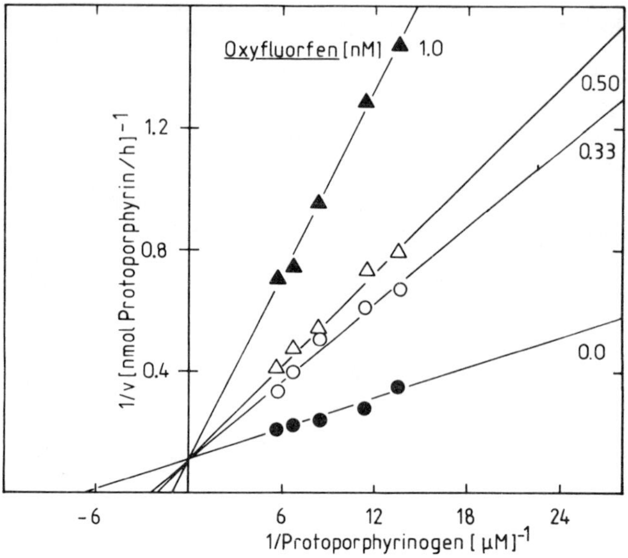

Figure 3. Initial inhibition of maize protoporphyrinogen oxidase by several concentrations of oxyfluorfen (indicated in n*M* at the regression lines) at various protoporphyrinogen concentrations in crude etioplast preparations under standard assay conditions. Nonenzymatic conversion of protoporphyrinogen by heat-inactivated enzyme was subtracted separately at each substrate concentration. The regression coefficient was better than 0.97.

formation up to 30 μg chlorophyll per milliliter of assay mixture. Therefore, plastids from greening plants should be used to achieve the best enzyme activity to chlorophyll ratio. The sensitivity of protoporphyrin detection can be enhanced by the addition of a detergent (Tween 80) to the assay mixture. Protoporphyrinogen oxidase activity determination is influenced by the unspecific chemical oxidation of protoporphyrinogen resulting in nonlinear rates of protoporphyrin IX formation. To minimize the nonenzymatic oxidation of protoporphyrinogen a reductant should be added to the reaction mixture. For maize protoporphyrinogen oxidase, DTT was most suitable because the autoxidation rate was strongly suppressed while the enzyme activity was only slightly affected. For each source of enzyme the optimum reducing agent and concentration should be determined separately.

ACKNOWLEDGMENT

Our studies were supported by the Fonds der Chemischen Industrie.

REFERENCES

Matringe, M., Camadro, J. M., Labbe, P., and Scalla, R., Protoporphyrinogen oxidase as a molecular target for diphenyl ether herbicides. *Biochem. J.*, 260, 231, 1989.

Sandmann, G. and Böger, P., Accumulation of protoporphyrin IX in the presence of peroxidizing herbicides. *Z. Naturforsch.*, 43c, 699, 1988.

Becceril, J. M. and Duke, S. O., Protoporphyrin IX content correlates with activity of photobleaching herbicides. *Plant Physiol.*, 90, 1175, 1989.

Labbe, P., Camadro, J. M., and Chambon, H., Fluorimetric assays for coproporphyrinogen oxidase and protoporphyrinogen oxidase. *Anal. Biochem.*, 149, 248, 1985.

Jacobs, N. J. and Jacobs, J. M., Assay for enzymatic protoporphyrinogen oxidation, a late step in heme synthesis. *Enzyme*, 28, 206, 1982.

CHAPTER 7

Short-Term Formation of Protoporphyrin IX in *Scenedesmus* Using Cyclic Imide Analogues

Hiroyuki Watanabe, Gerhard Sandmann, Ko Wakabayashi, and Peter Böger

INTRODUCTION

Peroxidizing herbicides such as *p*-nitrodiphenyl ethers, cyclic imides, oxadiazoles, or pyrazole derivatives inhibit chlorophyll biosynthesis and induce lipid peroxidation, disruption of membranes, chlorophyll degradation, and desiccation in the light. They inhibit protoprophyrinogen oxidase, which leads to rapid accumulation of protoporphyrin IX (proto-IX), an intermediate of chlorophyll biosynthesis. Abnormal accumulation of proto-IX in plant tissue is assumed to cause light-induced formation of radicals which initiate the peroxidative reaction described (see Böger and Sandmann, 1990, for review).

Our study presents a simple method to determine the trigger substance, proto-IX, using the unicelluar green alga *Scenedesmus acutus*. Having examined various cyclic imide analogues, it was found that the *short-term* formation of proto-IX represents a reliable and quantitative indicator for phytotoxicity of peroxidizing herbicides.

0-87371-539-X/93/$0.00 + $.50

PROCEDURE

Algal Cultivation

See Chapter 8 for autotrophic cultivation of *Scenedesmus acutus* (strain no. 276-3a of the Algae Collection, University of Göttingen, D-3400 Göttingen, Germany). Light intensity is 80 μEinstein (E)/m^2 × sec, which is about 6000 lux.

Protoporphyrin-IX Determination

After a 24-hr autotrophic growth period the peroxidizing herbicide is added; proto IX is then determined after a 1-hr treatment or at the time indicated. The herbicide is dissolved in ethanol, keeping the final ethanol content below 0.1% (v/v) in the culture medium. A 50-mL volume of the cell suspension is quickly centrifuged at 2500 × *g* for 5 min at about 6°C in the dark; approximately 250 μL packed cell volume (pcv) are used for proto-IX determination.

Extraction and determination of proto-IX is performed according to the method of Ho et al. (1987) with some modification. After washing twice with 25 mL of ice-cold distilled water, the cells are extracted twice with 2 mL of the extraction mixture containing methanol, tetrahydrofuran, and a 5 m*M* aqueous solution of trifluoroacetic acid (30:16:5, v/v/v) at 55°C for 10 min with shaking in the dark. After removal of the precipitate by centrifugation, proto-IX in the dark-green extract is determined by high-pressure liquid chromatography (HPLC) separation with a fluorescence detector (Hitachi D-6000 system). The column used is Senshu Pak C_6H_5-1252-N (250 × 4.5 mm inner diameter; Senshu Scientific Co., Tokyo, Japan) with a Guard Pak CN precolumn (Waters Associates, Milford, Connecticut). The eluent is a mixture of methanol, tetrahydrofuran, and a 5 m*M* aqueous solution of trifluoroacetic acid (24:13:13, v/v/v) with a flow rate of 1 mL/min. Proto-IX concentration is measured by fluorescence using the excitation wavelength of 405 nm and the emission wavelength of 633 nm. For calibration, an authentic proto-IX standard from Sigma Chemical Co., St. Louis, Missouri is used.

Chlorophyll Determination

Total chlorophyll content is determined by the measurement of optical density of the methanolic extract at 650 and 665 nm. A 2-mL volume of the cell suspension is centrifuged at 2500 × *g* for 5 min and washed once with ice-cold distilled water. Chlorophyll is extracted by incubation of the pellet with 5 mL of methanol at 65°C for 10 min. The total amount of chlorophyll is calculated by the following equation:

$$\text{Total chlorophyll } (\mu\text{g/mL algal suspension}) = (25.5 \times OD_{650} + 4.0 \times OD_{665}) \times C$$

where C = dilution (volume of methanol/algal cell suspension; C = 2.5 in this experiment

Equipment

HPLC equipment for the proto-IX determination is the Hitachi D-6000 system with an F-1050 fluorescence spectrophotometer, L-4200 UV-VIS detector, L-6200 intelligent pump, 655A-40 autosampler, D-6000 HPLC manager, and interface. Microcentrifuge tubes for the measurement of packed cell volume were purchased from Kummer, D-7500 Freiburg, Germany. The illuminated and thermostated growth water bath and cell culture tubes were obtained from Shimomura Scientific Co., Tokyo, Japan. Suitable thermostated water baths are also supplied by Kniese-Edwards, D-3550 Marburg, Germany.

Herbicides

N-substituted 3,4,5,6-tetrahydrophthalimides were prepared by the condensation reaction of 3,4,5,6-tetrahydrophthalic anhydride and appropriate amines according to the method given elsewhere (Ohta et al., 1980; see Watanabe et al. (1992) for further details on synthesis).

RESULTS

Comparison of Acid and Alkaline Extraction for Proto-IX Determination

At first, the acid and alkaline extraction systems of proto-IX were compared because some laboratories studying photobleaching herbicides use an alkaline medium for proto-IX extraction (e.g., Matringe and Scalla, 1988).

After treatment with 1 μ*M* chlorophthalim, *N*-(4-chlorophenyl)-3,4,5,6-tetrahydrophthalimide, *Scenedesmus* cells are washed twice with ice-cold distilled water and proto-IX is extracted with the acid extraction mixture described in the procedure section. For comparison, an alkaline solution was used; this was a mixture of acetone and a 0.1 *M* aqueous solution of NH_4OH (9:1, v/v). A total of 10 μL of each extract is injected into the HPLC system as described. A chromatogram is shown in Figure 1. As shown in Table 1, the acid extract contains more proto-IX than the alkaline extract. Thus, the acid extraction is used for the following experiments to determine the proto-IX content of *Scenedesmus* cells.

Time and Dose Dependency of Proto-IX Formation

Time-dependent changes in proto-IX and chlorophyll content are shown in Figure 2. Proto-IX accumulates in the cells after chlorophthalim application (0.1 to 10 μ*M*) and reaches the maximum level after 5 to 10 hr. The initial

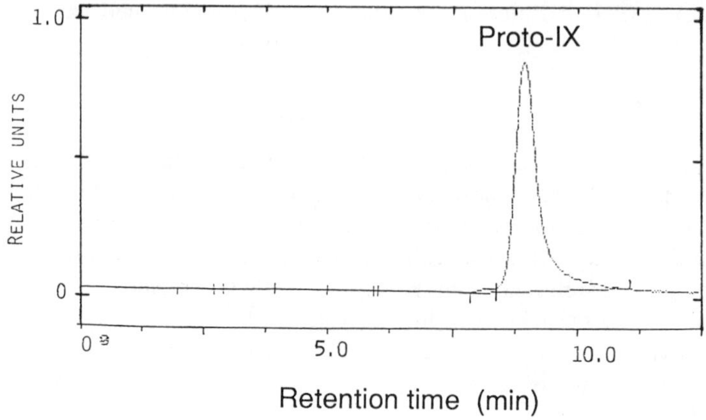

Figure 1. Chromatogram of protoporphyrin IX from autotrophic *Scenedesmus.* Proto-IX was obtained by acid extraction 1 hr after incubation with 1 μ*M* chlorophthalim. The peak area is equivalent to 65.2 nmol proto-IX per milliliter of pcv; retention time = 9.1 min. When fractionating the *alkaline* proto-IX extract (see Proto-IX determination section in Procedure), retention time is prolonged to about 14 min. Spectral analysis indicates that proto-IX is essentially identical using both procedures. Apparently, the ammonium salt formed by the alkaline extraction is more lipophilic than the acidic proto-IX form.

velocity of proto-IX formation clearly depends on herbicide concentration, but apparently the maximum level attained eventually is not dependent on concentration. After the maximum accumulation, proto-IX rapidly decreases (seemingly by photo-oxidation) concurrently with chlorophyll degradation.

Figure 3 shows the dose effect of chlorophthalim on proto-IX accumulation for exactly 1 hr after the application. The 1-hr incubation is the minimum period in which to take samples and to reliably determine the proto-IX content. Using a 1-hr treatment, 16 nmol proto-IX per 250 μL pcv are obtained by acid extraction, while the control contains 0.2 to 0.3 nmol per 250 μL pcv. It could be demonstrated that the *short-term* accumulation of proto-IX just after herbicide application is related to herbicide activity of cyclic imide analogues at a given concentration.

Table 1. Comparison of Acid and Alkaline Extraction for Protoporphyrin IX Determination

Extraction system	Proto-IX (nmol/mL packed cell volume)[a] Herbicide treatment for: 1 hr	3 hr	5 hr
Acid extract	66.7 ± 4.1	153.3 ± 11.1	181.9 ± 10.3
Alkaline extract	48.3 ± 2.8	114.7 ± 10.8	129.7 ± 9.5

[a] Values are shown as means ± SD of six determinations in three separate experiments.

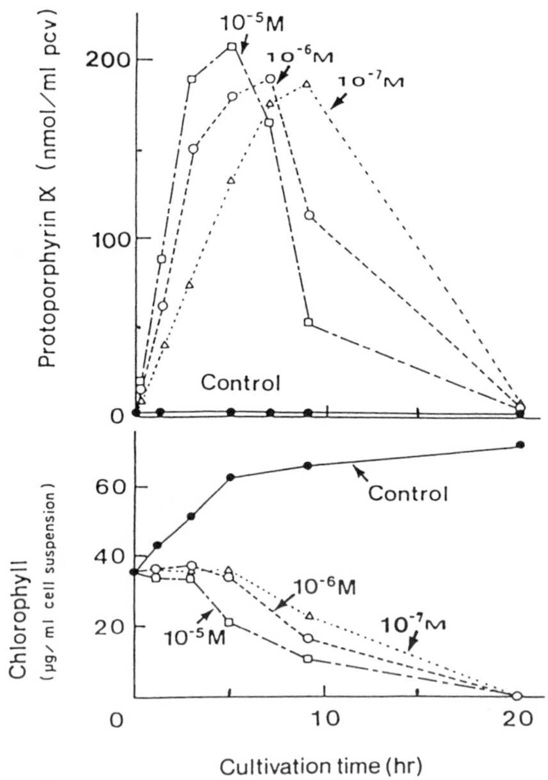

Figure 2. Time-dependent changes in protoporphyrin IX and chlorophyll content in autotrophic *Scenedesmus*. Cells were incubated with chlorophthalim using the concentrations indicated. Note that after about 5 to 6 hr proto-IX decreases, as does chlorophyll, apparently due to photo-oxidation. The bottom curves also demonstrate that peroxidizing herbicides cause an immediate halt of chlorophyll biosynthesis.

COMMENTS

1. Abnormal accumulation of proto-IX in plant is assumed to be a primary trigger for the photobleaching effect of cyclic imides, diphenyl ethers, and other peroxidizing herbicides. Therefore, quantitative determination of proto-IX accumulation is an indicator to evaluate the photobleaching and herbicidal activity of such compounds. Proto-IX formation in the plant strongly depends on the activity of chlorophyll biosynthesis, growth stage, and growth condition of the plant. Accordingly, a *Scenedesmus* culture of the logarithmic growth phase should be used. A mixotrophic (photoheterotrophic) culture is even stronger in proto-IX accumulation. In this procedure, by using a unicellular green algae species, formation of proto-IX is simply and quantitatively determined with good reproduc-

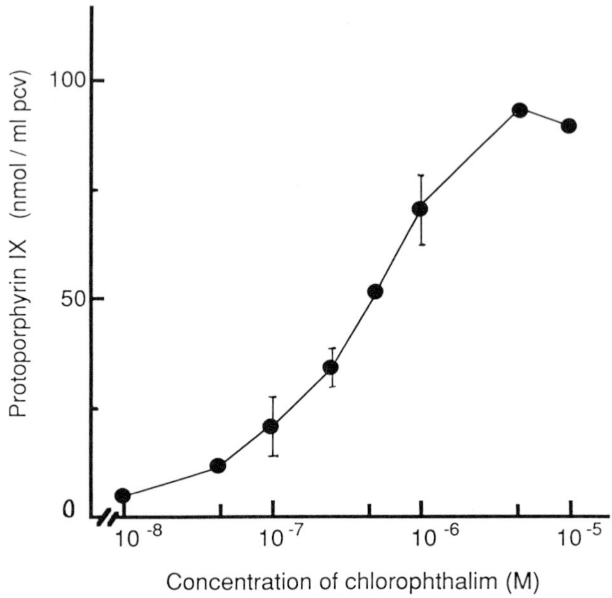

Figure 3. Dose effect of chlorophthalim on the 1-hr formation of protoporphyrin IX in autotrophic *Scenedesmus.* Proto-IX was extracted exactly 1 hr after administration of chlorophthalim. Points with error bars show the values as means ± SD of two or three separate experiments with duplicate assays.

ibility. A quantitative measurement with higher plants (e.g., seedlings) is more complicated.

2. The short-term proto-IX accumulation by 17 cyclic imide compounds and the diphenyl ether oxyfluorfen was measured in *Scenedesmus* cells exactly 1 hr after incubation with each compound at 1 μM concentration. Bleaching of chlorophyll, cell growth inhibition, and herbicidal activity (root growth inhibition, and herbicidal activity (root growth inhibition of *Echinochloa utilis*) by these compounds correlate quantitatively with the 1-hr accumulation of proto-IX. The regression equations are given in Table 2. No quantitative correlation was found with *long-term* proto-IX accumulation (see Watanabe et al., 1992).

Table 2. Quantitative Correlation between Short-Term Protoporphyrin IX Formation and Other Phytotoxic Parameters[a]

(1) pI_{50} (chlorophyll) = 0.040 Proto-IX + 4.918
(±0.008) (±0.421)

n = 18, r = 0.936, F = 113.7, s = 0.495

(2) pI_{50} (growth) = 0.039 Proto-IX + 4.884
(±0.009) (±0.456)

n = 18, r = 0.920, F = 88.1, s = 0.536

(3) pI_{50} (herbicidal activity) = 0.0032 Proto-IX + 3.830
(±0.013) (±0.701)

n = 18, r = 0.784, F = 25.5, s = 0.824

[a] Compiled after data in Watanabe et al. (1992): (1) and (2) are from autotrophic *Scenedesmus acutus* cultures; (3) is the inhibition of root growth of sawa millet (*Echinochloa utilis*). Data in parentheses are the 95% confidence intervals.

REFERENCES

Böger, P. and Sandmann, G., Modern herbicides affecting typical plant processes, in *Chemistry of Plant Protection,* Vol. 6, Bowers, W. S., Ebing, W., Martin, D., and Wegler, R., Eds., Springer-Verlag, Berlin, 1990, p. 173.

Ho, J., Guthrie, R. and Tieckelmann, H., Quantitative determination of porphyrins, their precursors and zinc protoporphyrin in whole blood and dried blood by high-performance liquid chromatography with fluorometric detection. *J. Chromatogr. (Biomed. Appl.),* 417, 269, 1987.

Matringe, M. and Scalla, R., Studies on the mode of action of acifluorfen-methyl in nonchlorophyllous soybean cells. *Plant Physiol.,* 86, 619, 1988.

Ohta, H., Jikihara, T., Wakabayashi, K., and Fujita, T., Quantitative structure-activity study of herbicidal N-acryl-3,4,5,6-tetrahydrophthalimides and related cyclic imides. *Pestic. Biochem. Physiol.,* 14, 153, 1980.

Watanabe, H., Ohori, Y., Sandmann, G., Wakabayashi, K., and Böger, P., Quantitative correlation between short-term accumulation of protoporphyrin IX and peroxidative activity of cyclic imides. *Pestic. Biochem. Physiol.,* 42, 99, 1992.

CHAPTER 8

Ethane Formation by Peroxidizing Herbicides

Peter Böger and Beate Nicolaus

INTRODUCTION

Radical-producing herbicides like *p*-nitrodiphenyl ethers, cyclic imides, and others (Figure 1) cause formation of short-chain hydrocarbons (see Sandmann and Böger, 1990, for review and possible reaction sequences). These gases originate from polyunsaturated fatty acids being constituents of acyl lipids in the cell and chloroplast membranes. Short-chain hydrocarbons produced are mostly alkanes, with some unsaturated alkenes, and exhibit a chain length of ω-1 (Sandman and Böger, 1982). "ω" indicates the location of a double bond closest to the reduced side of the fatty acid. Since α-linolenic acid is the prominent fatty acid in eukaryotic thylakoids, ethane is the major gas observed. Although ethane is also produced with bipyridylium herbicides present, interest in this assay appeared with the development of protoporphyrinogen oxidase inhibitors (peroxidizing compounds), which leads to overproduction of protoporphyrin IX in the cell (Matringe and Scalla, 1988). This tetrapyrrole is sensitized by light and radicals are produced, which cause degradative reactions.

Using *Scenedesmus* cultures, ethane formation induced by peroxidizers can be *quantitatively* correlated with other cell data like growth impairment or chlorophyll bleaching. These figures again can be correlated with herbicidal

0-87371-539-X/93/$0.00 + $.50

PEROXIDIZING HERBICIDES

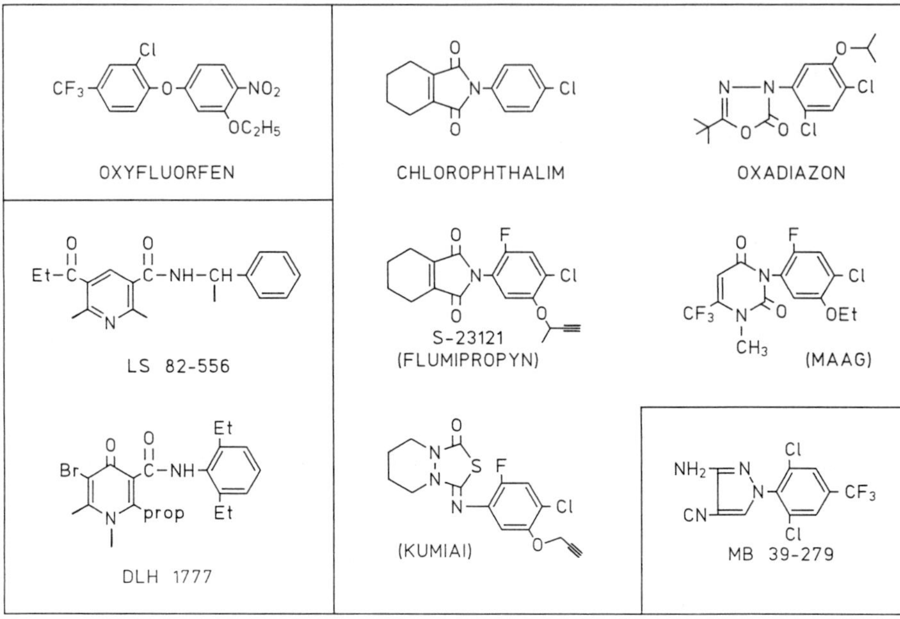

Figure 1. Formulas of so-called "peroxidizers" which induce accumulation of tetrapyrroles (mostly protoporphyrin IX) in plastid-containing cells.

greenhouse effects on weeds or with a laboratory root-tip growth assay (Wakabayashi et al., 1988; Chapter 7). Recently it could be shown that ethane production of *Scenedesmus* significantly correlates with inhibition of protoporphyrinogen oxidase from corn (Figure 2 and Chapter 6). Accordingly, the ethane assay as described here is a fast and convenient routine method for both activity screening and obtaining data for structure/activity considerations.

The action of peroxidizing herbicides is most apparent in the light with the chloroplast as the site of attack. Using autotrophic cells, photosynthesis inhibitors present additionally alleviate peroxidizing activity (Nicolaus et al., 1989). Protoporphyrinogen oxidase inhibitors can also be inhibited by gabaculine (2-amino-2,3-dihydrobenzoate) or 2,4-dioxoheptanoic acid (Duke et al., 1991). Such inhibitors should be included as controls when peroxidative ethane formation is measured.

Soybean cell cultures also can be used to determine activity of peroxidizing herbicides, although the quantitative reliability is not so good. Like algal cells, both autotrophic and glucose-supplemented culture conditions can be established. Cultivation, however, is more complicated (Sandmann et al., 1989).

PROCEDURE

Growth of *Scenedesmus acutus*

The strain is no. 276-3a of the Algae Collection, University of Göttingen, D-3400 Göttingen, Germany. The mineral medium, as given in Table 1, is sterilized, inoculated with a cell density of 1 μL packed cell volume (pcv) per milliliter of culture suspension, and grown for about 24 hr to attain a density of about 2 μL pcv per milliliter. Sterile cultivation is done in a thermostated waterbath at 22°C in closed glass culture vessels (4 × 40 cm) which can be gassed with sterile CO_2-enriched air (4% v/v). Gassing is through an inlet reaching to the bottom of the vessel. The gassing stirs the suspension and prevents sedimentation. Light intensity is about 80 μEinstein (μE)/m^2 × sec, equivalent to about 6000 lux; normal fluorescent light is sufficient. For (dark) heterotrophic growth the culture is inoculated with 2 μL pcv/per milliliter of suspension; 3 to 4 μL pcv per milliliter is attained after a 24-hr cultivation period. Dark (green) cultures can be used advantageously to investigate herbicide-induced protoporphyrin IX formation since photo-oxidative degradation of the tetrapyrrole is avoided.

Incubation for Ethane Formation

A 2-ml volume of algae suspension (autotrophic cells, 2 μL pcv per milliliter; heterotrophic ones, 3 μL pcv per milliliter) are shaken in a thermostated Warburg apparatus for 17 hr (generally overnight) at 23°C in 9.5-mL vials

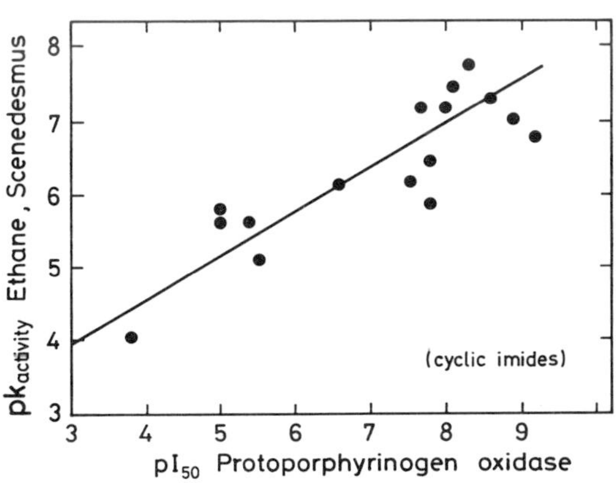

Figure 2. Correlation of light-induced ethane formation by autotrophic *Scenedesmus acutus* (logarithmic activity values, $pk_{activity}$) with inhibition of protoporphyrinogen oxidase from corn etioplasts (pI_{50} values). Modified cyclic imides with a broad range of peroxidative activity were used; each point refers to an analogue. For measurement of protoporphyrinogen oxidase activity see Chapter 6. The regression equation is highly significant. (After Nicolaus et al., unpublished results of our laboratory.)

Table 1. Medium for *Scenedesmus* Cultures

Stock Solutions for Mineral Medium[a]				Milliliters of Stock Used per Liter of Culture Medium
Stock I:	100	g	KNO_3	
	26	g	$Na_2HPO_4 \cdot 2\ H_2O$	
	74	g	KH_2PO_4	
	5	g	$MgSO_4 \cdot 7\ H_2O$	
	Fill up to 100 mL			10
Stock II:	1.89 $CaCl_2$			
	Fill up to 250 mL			1
Stock III:	2.2	mg	$Al_2(SO_4)_3 \cdot 18\ H_2O$	
	33.0	mg	$MnCl_2 \cdot 4\ H_2O$	
	10.0	mg	$CuSO_4 \cdot 5\ H_2O$[b]	
	0.93	mg	$CoSO_4 \cdot 7\ H_2O$	
	2.1	mg	$ZnSO_4 \cdot 7\ H_2O$	
	10.2	mg	H_3BO_3	
	0.6	mg	$(NH_4)_6Mo_7O_{24} \cdot 4\ H_2O$	
	0.97	mg	NH_4VO_3	
	8.76	mg	$NiSO_4 \cdot 6\ H_2O$	
	83.0	mg	KI	
	79.0	mg	KBr	
	2.8	mg	LiCl	
	Fill up to 1000 mL			3
	Keep in refrigerator			
Stock IV:	1.45	g	$Fe(NO_3)_3 \cdot 9\ H_2O$	
	1.33	g	EDTA	
	Fill up to 500 mL			5

[a] Use distilled water; pH of complete medium: 6.4. For glucose-supplemented growth (heterotrophic or mixotrophic) add 0.5% glucose and 0.25% yeast extract (Difco). The complete culture medium is sterilized. Keep stock cultures on agar slants prepared with mineral medium + 2% Difco agar. Use screw-capped tubes and store at room temperature and in dim light. Renew stock cultures every year.

[b] May vary between 2.5 and 30 μg/L.

sealed with rubber septa. Light intensity is 350 to 400 $\mu E/m^2 \times sec$ or about 20,000 lux, supplied by tungsten lamps. Bicarbonate (5 m*M*) is added to autotrophic cells to ensure photosynthesis at the start of the incubation (pH 7.2). All assays are run in duplicate. Gas samples (400 μL) are withdrawn by a head-space injector HS6, mounted on a Perkin-Elmer gas chromatograph F22, and determined by flame ionization after separation on an activated alumina column (temperature program from 70 to 140°C).

Chlorophyll is extracted with 60°C methanol and determined in a spectrophotometer. Packed cell volume is determined in simple gradulated microcentrifuge tubes of 80 μL capacity (see Chapter 7). Herbicides are dissolved in methanol, acetone, or dimethylsulfoxide (DMSO); the solvent should be kept below 0.1%, v/v.

Growth of Soybean Cells in Liquid Culture

The medium, as given in Table 2, is inoculated with about 20 to 30 μL pcv per milliliter of culture suspension and grown of 8 to 12 days in double Erlenmeyer flasks (Table 3, Figure 3). Light incubation for gas production

Table 2. Medium for Soybean Cell Cultures[a]

Murashige-Skoog medium[b]	4.71 g/L
Vitamins[c]	1 mL/L
Saccharose	10 g/L
NAA solution[d]	1 mL/L
Kinetin solution[e]	1 mL/L
Adjust to pH 6.2	
Agar (for petri dish stock cultures)	3 g/500 mL

[a] For mixotrophic and autotrophic cultivation. Omit saccharose in the latter case. The complete medium is sterilized.

[b] Murashige-Skoog medium: Bought as a complete dry mixture from, e.g., Flow Laboratories, D-5309 Meckenheim, Germany.

[c] Vitamins (prepared with sterile distilled water): nicotinamide (vitamin B_5), 0.2 mg/L; pyridoxin·HCl (vitamin B_6), 0.2 mg/L; *d*-biotin, 0.1 mg/L; choline chloride, 0.1 mg/L; Ca-pantothenate, 0.1 mg/L; thiamine·HCl (vitamin B_1), 0.1 mg/L; folic acid, 0.05 mg/L; *p*-aminobenzoic acid, 0.05 mg/L; riboflavin, 0.05 mg/L; cyanocobalamine (vitamin B_{12}), 0.00015 mg/L. Prepare 1-mL samples in sterile vials; store in deep freeze.

[d] NAA solution (store in refrigerator): α-naphthylacetic acid (e.g., from Serva, Heidelberg, Germany), 10 mg/10 mL ethanol p.a.

[e] Kinetin solution (store in refrigerator): Dissolve 20 mg kinetin in approximately 1 mL of ethanol and bring volume up to 100 mL with distilled water.

is performed as described for *Scenedesmus*. Cell density in the vials is 15 to 30 μL pcv per milliliter (or more) for autotrophic cells and 40 to 60 μL pcv per milliliter for mixotrophic cells, suspended in their nutrient media.

RESULTS

Peroxidation with *Scenedesmus*

Table 4 shows typical data of an incubation experiment with three cyclic imide analogues of different peroxidative activity. Gas formation is measured

Table 3. Growth of Soybean Cells

	Autotrophic		Mixotrophic	
Cultivation Time	μL pcv / mL[b]	μg Chl[a] / mL pcv[c]	μL pcv / mL	μg Chl / mL pcv
Start	20	—	20	—
4 days	28	155	50	192
6 days	60	218	185	173
12 days	58	264	200	243

[a] Chl = chlorophyll.

[b] mL culture suspension.

[c] pcv = packed cell volume.

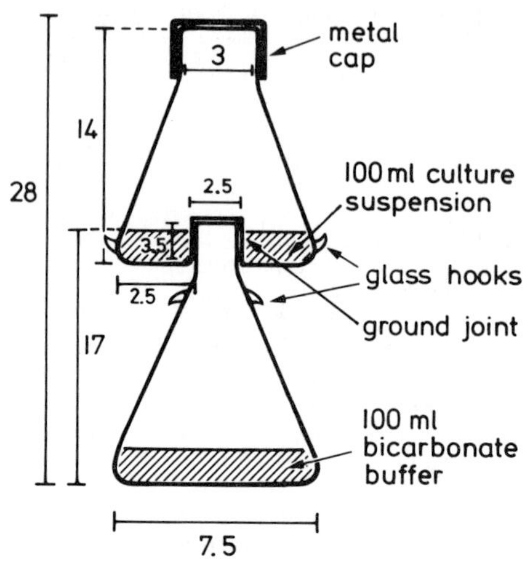

Figure 3. Cultivation of autotrophic soybean cells in "double flasks" made by Markus Co., D-7888 Rheinfelden, Germany. All figures are in centimeters. The culture is kept in a climate chamber at 25°C. Illumination is from the top with a light intensity of about 55 $\mu E/m^2 \times$ sec (3500 lux) at the cultivation suspension level. Light is provided by fluorescent lamps, which have a strong proportion of red light. The bicarbonate buffer (mixture of 8 g/L Na_2CO_3 and 35.7 g/L $NaHCO_3$) ensures CO_2 enrichment of the atmosphere in the cultivation vessel. The flasks are fixed by springs (see the glass hooks) and are placed on a shaker (115 rpm) at 25°C. The mixotrophic culture is grown in simple Erlenmeyer flasks of 250 mL volume, also using 100 mL of culture suspension per vessel. See Table 3 for inoculum.

for some concentrations in the range which will induce half the maximum gas formation; it is referred to milliliters of cell suspension and milliliters of pcv (last two columns). The data of the first three columns are internal controls. As shown by the second and third columns, chlorophyll per milliliter and chlorophyll per pcv decrease (photobleaching), no change vs. start conditions is observed with a peroxidizer + diuron present (fourth line). Chlorophyll per milliliter of the control (second line) increases during the 17-hr incubation time. The slight increase of pcv per milliliter (first column) apparently is due to an increase in cell volume, which may be caused by disturbed cell permeability. Deviation from the mean is about ±10%. Two separate sets of experiments may differ 20 to 30%; the herbicide-induced gas production pattern, however, is always the same. Gas formation down to 1.8 pmol/mL algae suspension can be reliably determined. Gases formed are ethane and ethylene; the latter is about 15% of the total gas production (Sandmann and Böger, 1990). The control is 0.4 nmol C_2 gas per milliliter pcv or less.

In Figure 4 the data of Table 3 are shown as double-reciprocal plots to allow for mathematical calculation of "activity values" by regression line equations. A common intersection of the ordinate shows up, indicative of *one*

Table 4. Light-Induced Short-Chain Hydrocarbon Formation by Autotrophic *Scenedesmus* Under the Influence of Peroxidizing Compounds

Additions		μl pcv / mL	μg Chl / mL	mg Chl / mL pcv	Ethane + Ethylene nmol / mL	Ethane + Ethylene nmol / mL pcv
Start of light incubation		2.0	21.8	10.9	—	—
Control		3.4	25.9	7.5	1.7	0.4
Oxyfluorfen, 10^{-6} *M*		2.6	11.0	4.3	28.0	11.0
Oxyfluorfen, 10^{-6} *M* + diuron, 10^{-5} *M*		2.6	20.0	10.0	2.0	1.0
Compound 1:	7×10^{-6} *M*	2.7	20.8	7.7	5.9	2.2
	10^{-5} *M*	2.9	19.9	6.9	8.0	2.8
	3×10^{-5} *M*	2.5	18.2	7.3	17.4	7.0
	5×10^{-5} *M*	2.6	17.3	6.6	12.8	4.9
	7×10^{-5} *M*	2.1	20.7	9.9	14.7	7.0
	10^{-4} *M*	2.4	14.4	6.0	17.9	7.5
Compound 2:	8×10^{-8} *M*	2.9	16.5	5.7	8.7	3.0
	10^{-7} *M*	2.9	14.9	5.1	10.4	3.6
	1.25×10^{-7} *M*	2.9	13.3	4.6	11.9	4.1
	1.5×10^{-7} *M*	2.9	10.7	3.7	14.2	4.9
	2×10^{-7} *M*	2.9	8.7	3.0	15.7	5.4
Compound 3:	6×10^{-6} *M*	3.1	21.0	6.8	11.8	3.8
	7×10^{-6} *M*	3.2	21.1	6.6	13.4	4.2
	10^{-5} *M*	3.1	14.6	4.7	17.4	5.6
	1.5×10^{-5} *M*	3.0	8.0	2.7	21.3	7.1
	5×10^{-5} *M*	2.9	10.9	3.7	22.8	7.9

Note: Compounds 1, 2, and 3 are cyclic imide analogues (see Figure 1); mL refers to algal suspension; mL pcv denotes mL packed cell volume. When the assay samples are kept in the dark neither a change in chlorophyll (Chl) content vs. control nor gas production is observed.

common and decisive reaction site, although ethane is an end product of a longer reaction sequence. It is noteworthy that these activity values correlate significantly with I_{50} values of protoporphyrinogen oxidase from corn (Figure 2). The enzyme of *Scenedesmus* is also inhibited correspondingly (unpublished results of our laboratory).

Peroxidation with Soybean Cells

As shown by Figure 5, ethane, propane, and ethylene are produced in the light with a peroxidizing herbicide (chlorophthalim) present. All gases are due to the peroxidative herbicide effect, although part of the ethylene is a stress response shown by higher plant cells only. (This ethylene does not originate from α-linolenic acid and can be suppressed by inhibitors of aminocyclopropanecarboxylic acid synthase.) Diuron blocks alkane formation in autotrophic soybean cells only; no effect is seen with mixotrophic cultures. Lack of diuron inhibition has also been shown with glucose-supplemented (heterotrophic or mixotrophic) *Scenedesmus* (Nicolaus et al., 1989). In such cells photosynthesis is not required to accumulate protoprophyrin IX. Include gabaculine or 2,4-dioxoheptanoic acid (1 m*M*) as a control inhibitor to block formation of short-chain alkanes.

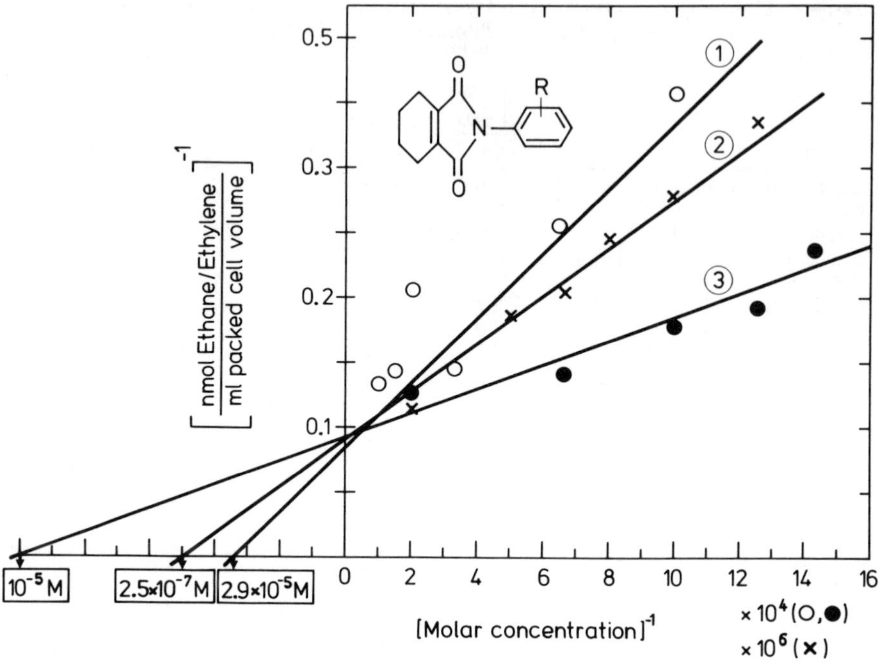

Figure 4. Determination of light-induced peroxidative activity of three cyclic imide analogues (nos. 1, 2, and 3). Ethane (and some ethylene) produced by autotrophic *Scenedesmus acutus* is plotted against different concentrations of the imides. This double-reciprocal plots yield the "activity values" (indicated in boxes), which represent the concentration of the imides needed to obtain half of the maximum ethane formation. The original data of this experiment are given in Table 4.

COMMENTS

1. To determine peroxidative alkane formation, vigorously growing cells are advantageous for high yields. For *Scenedesmus,* ethane + ethylene should be 10 to 15 nmol/mL pcv, not less. The diuron inhibition may be poor in autotrophic cells, when too much starch is accumulated, but the reliability of the experiment should not be influenced. It is recommended to start the liquid culture from the agar slant after 2 to 3 months of continuous growth in case the required yields are no longer attained.
2. Soybean cells do not grow as rapidly as *Scendesmus.* The yield of herbicide-induced hydrocarbons is comparatively low (see Figure 5), although this disadvantage can be overcome by incubating denser cell suspensions. The ethylene, which is about 50% of the two alkanes formed, should be subtracted when activity values are of interest.
3. We consider these two systems as appropriate with preference to *Scenedesmus.* Cress seedlings (*Lepidium sativum*) can also be used. Let 30

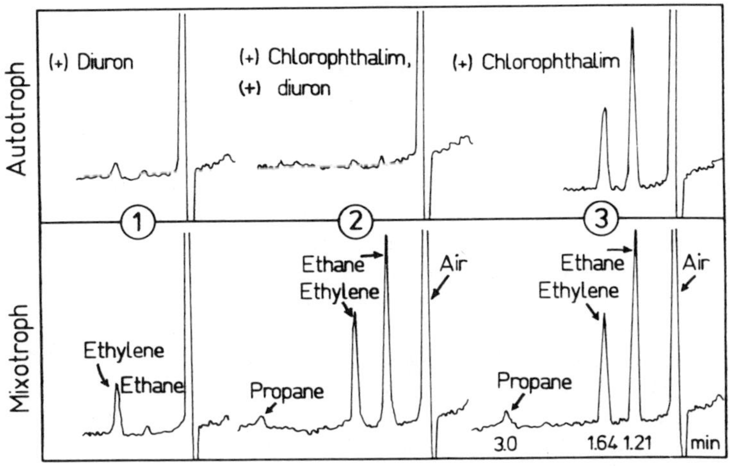

Figure 5. Light-induced production of short-chain hydrocarbons by soybean cell cultures (*Glycine max.* Merr.) with 0.5 μM chlorophthalim present. Cells were grown either autotrophically or mixotrophically (photoheterotrophically). Diuron (10 μM) inhibited gas formation (and accumulation of protoporphyrin IX) in autotrophic cells only (see Sandmann et al., 1989). A total of 2.5 to 3.5 nmol short-chain hydrocarbons per milliliter packed cell volume is produced during a 17-hr assay. Retention times for the three hydrocarbon gases are given in No. 3, bottom.

seeds germinate on two layers of wet filter paper in petri dishes for 3 days in the dark at about 22°C, then illuminate them for 5 hr with about 12,000 lux and place 20 seedlings into the incubation vials as used in the section on incubation for ethane formation in the procedure section, with 3 mL of 5 m*M* phosphate buffer, pH 7.5. Incubation is for 16 to 17 hr at 12,000 lux (about 190 $\mu E/m^2 \times sec$) fluorescent light, at 22°C, with gentle shaking (160 rpm). Besides ethylene, ethane and propane are also formed. The yield of the two alkanes is up to 10 pmol per seedling, while ethylene is always higher. The simple assay system allows to check — preferably with a 10 μM concentration in triplicate — whether new compounds are active or not. This information is often sufficient for chemists, particularly since a higher plant is used. The cress seedlings system — as it is worked out at the moment — is not recommended for a quantitative evaluation as done in Figure 2, since the deviation of gas formation between samples is quite high (20 to 50%). Most probably this is due to the unavoidable handling of the seedlings before the incubation experiment.

ACKNOWLEDGMENT

This study was supported by the Fonds der Chemischen Industrie.

REFERENCES

Duke, S. O., Becerril, J. M., Sherman, T. D., and Matsumoto, H., Photosensitizing porphyrins as herbicides, in *Naturally Occurring Pest Bioregulators,* Hedin, P. A., Ed., ACS Symp. Ser. 449, American Chemical Society, Washington, D.C., 1991, p. 371.

Matringe, M. and Scalla, R., Studies on the mode of action of acifluorfen-methyl in nonchlorophyllous soybean cells. *Plant Physiol.,* 86, 619, 1988.

Nicolaus, B., Sandmann, G., Watanabe, H., Wakabayashi, K., and Böger, P., Herbicide-induced peroxidation: influence of light and diuron on protoporphyrin IX formation. *Pestic. Biochem. Physiol.,* 35, 192, 1989.

Sandmann, G. and Böger, P., Peroxidizing herbicides, in *Managing Resistance to Agrochemicals,* Green, M. B., LeBaron, H. M., and Moberg, W. K., Eds., ACS Symp. Ser. 421, American Chemical Society, Washington, D.C., 1990, p. 407.

Sandmann, G. and Böger, P., Volatile hydrocarbons from photosynthetic membranes containing different fatty acids. *Lipids,* 17, 35, 1982.

Sandmann, G., Nicolaus, B., and Böger, P., Typical peroxidative parameters verified with mung-bean seedlings, soybean cells and duckweed. *Z. Naturforsch.,* 45c, 512, 1989.

Wakabayashi, K., Sandmann, G., Ohta, H., and Böger, P., Peroxidizing herbicides: comparison of dark and light effects. *J. Pestic. Sci.,* 13, 461, 1988.

CHAPTER 9

Peroxidizing Activity Determined by Cellular Leakage*

Stephen O. Duke and William H. Kenyon

INTRODUCTION

Eventually, any process that causes damage to the plasmalemma of a plant cell will cause the contents of the cell to leak. This process can occur without rupture of the plasmalemma; however, it is more pronounced when the plasmalemma is ruptured and the entire cell contents (both cytoplasmic and vacuolar) are released. Measurement of cellular leakage has been utilized to determine many types of stress to plants, including herbicidal damage (Duke et al., 1989; Matsumoto and Duke, 1990; Nandihalli et al., 1991; Kenyon et al., 1985). Cellular leakage is one of the first symptoms of peroxidative membrane damage due to protoporphyrinogen oxidase-inhibiting herbicides such as acifluorfen.

Many different types of cellular contents can be sampled to determine cellular leakage. One can measure leakage of specific endogenous compounds (e.g., amino acids or phenolic compounds; see, for example, Sato et al., 1988), fluorescent compounds, radiolabeled compounds that have been taken up by the cells, or endogenous electrolytes. Measurement of electrolytes offers a major advantage over the other methods. That is, leakage can be determined continuously with a conductivity probe and without destructive sampling. However, if the background conductivity of the bathing medium is high, measurement of change in conductivity may not be the method of choice for measuring cellular leakage.

Cellular leakage into the bathing media can be determined easily with cell cultures, floating leaf or cotyledon tissues (Duke et al., 1989; Kenyon et al., 1985), or intact aquatic plants (e.g., *Lemna* sp.; Matsumoto and Duke, 1990). The cellular damage of intact terrestrial plants may be determined by excising affected tissues, shaking them for a short time period in water or buffered solution, and then measuring the conductivity change in the wash solution (Nandihalli et al., 1991).

The specific procedure that is outlined in this chapter may be modified in many ways to suit the plant tissue, the herbicide, or the method of herbicide treatment. In this chapter, electrolyte leakage is used as the measure of cellular leakage caused by various peroxidizing herbicides.

PROCEDURE

Plant Material

This method can be adapted to any type of plant material. In the examples provided here, two types of plant tissues are used: 7-day-old cucumber cotyledons from which 4-mm discs were cut with a cork borer, and duckweed (*Lemna paucicostata*) fronds grown as previously described (Matsumoto and Duke, 1990). The cucumber cotyledon discs were washed in 1% sucrose, 1 m*M* 2-(morpholino)ethanesulfonic acid buffer (MES; pH 6.5), or 0.1% sucrose (no buffer) and then floated on 5 mL of the wash solution with or without herbicide in 6-cm petri dishes (50 discs per dish). Similarly, 50 colonies of duckweed were floated on 5 mL of distilled water, with or without herbicide, in 6-cm petri dishes.

Equipment

The only specialized equipment required is a conductivity meter and probe that will allow measurements within 1 μmho/cm. The experiments described here utilize a meter with a probe which allows the sampling of 1 mL of solution, which is returned after a measurement is made.

Herbicides

Technical-grade herbicides (>99%) and other chemicals were used in these studies. In some cases, the herbicides were dissolved in ethanol before adding them to the treatment solution. The ethanol concentrations in the cucumber and *L. paucicostata* media were never more than 0.5% and 0.3% (w/v), respectively, which caused no electrolyte leakage.

Methods of Expressing Data

The simplest method of expressing the data is to plot change in conductivity of treated tissue minus change in conductivity of the control with time. One can assume that the differential increase in cellular leakage is due to the treatment. An alternative is to plot the change in conductivity of both control and treated plant tissue. Plotting absolute values, instead of change in conductivity, can sometimes result in data that are difficult to compare because of initial differences in conductivity of the bathing solutions.

To compare effects of phytotoxins on different types of tissues or tissues from different plant species, normalized data, based on percent of total capacity to leak, should be utilized. Total capacity to leak can be determined by thoroughly homogenizing the tissues, centrifuging the homogenate, and determining the conductivity of the supernatant. This method has possible artifacts, in that hydrolytic enzymes may play different roles in generating electrolytes in homogenates and in tissues which are slowly losing their membrane integrity.

RESULTS

Typical electrolyte leakage kinetics due to peroxidative damage are shown for cucumber cotyledon tissue and duckweed colonies in Figure 1. Figure 1A illustrates the effectiveness of an inhibitor of porphyrin synthesis (gabaculine) in inhibiting cellular leakage caused by oxadiazon, an inhibitor of protoporphyrinogen oxidase. Figure 1B demonstrates the effectiveness of different acifluorfen-methyl concentrations on cellular leakage of *L. paucicostata* colonies. Note that the lag phase is independent of the herbicide concentration. There was little or no lag phase in the cucumber discs because the tissues were incubated in darkness for 20 hr before exposure to light (time 0 of the graph). High levels of protoporphyrin IX accumulated during this period (Duke et al., 1989). In *L. paucicosta,* 2 hr of exposure to light were required for relatively high levels of protoporphyrin IX to accumulate. The greater response of cotyledon discs than *L. paucicostata* plants to the herbicides is partly due to the greater tissue weight of the cucumber (ca. 0.2 g) vs. the duckweed (ca. 0.1 g). The more tissue per unit volume of bathing medium, the more sensitive the assay is.

Light-dependent or light-enhanced electrolyte leakage is common to a wide array of herbicide classes (Figure 2). However, the peroxidative damage due to inhibitors of protoporphyrinogen oxidase can be easily differentiated from that due to photosystem (PS) II inhibitors (e.g., diuron), PS I diverters (e.g., paraquat), or other herbicide classes by the protective effects of inhibitors of the porphyrin pathway such as gabaculine or 2,4-dioxoheptanoic acid (Figures 1A and 2). This is a quick and simple alternative to directly measuring effects on protoporphyrinogen oxidase. In cases in which a protoporphyrinogen oxidase inhibitor has a second site of action, the protective capacity of porphyrin synthesis inhibitors may be equivocal.

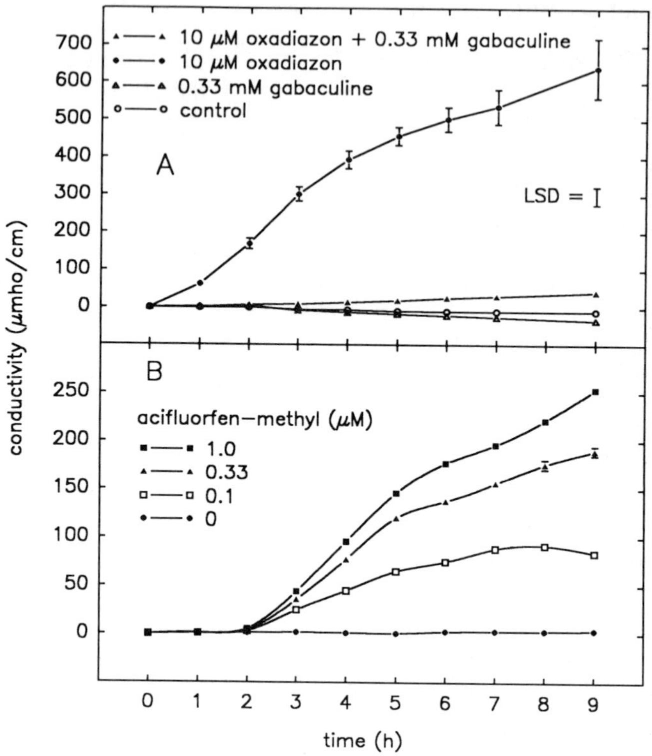

Figure 1. Increase in conductance of the bathing media of (A) cucumber cotyledon discs or (B) duckweed colonies after exposure to oxadiazon or acifluorfen-methyl, respectively. Cucumber tissues were incubated in darkness for 20 hr before exposure to light (time zero), whereas the duckweed was exposed to the herbicide and light simultaneously. Light in both experiments was 500 $\mu mol/cm^2/sec$). Error bars are ± 1 standard error of the mean. Error bars of some data points are obscured by the datum point symbol. Gabaculine is 2-amino-2,3-dihydrobenzoate; for chemical names of herbicides see Figure 2 legend. (Part A redrawn from Duke, S. O., Lydon, J., and Paul, R. N., *Weed Sci.,* 37, 152, 1989. With permission. Part B redrawn with permission from Matsumoto, H. and Duke, S. O., *J. Agric. Food Chem.,* 38, 2066, 1990. © 1990 American Chemical Society.)

COMMENTS

Although with this method results within an experiment are generally quite reproducible (see error bars in Figure 1), even small differences in tissue can greatly affect the magnitude of response with this assay. However, qualitative results are generally unaffected by tissue age, condition, etc. Treatment of cut leaf discs for long periods of time in solution may lead to artifacts due to growth of microbes in the external media. Use of buffers that are not good nutrients for microbes will minimize this potential problem.

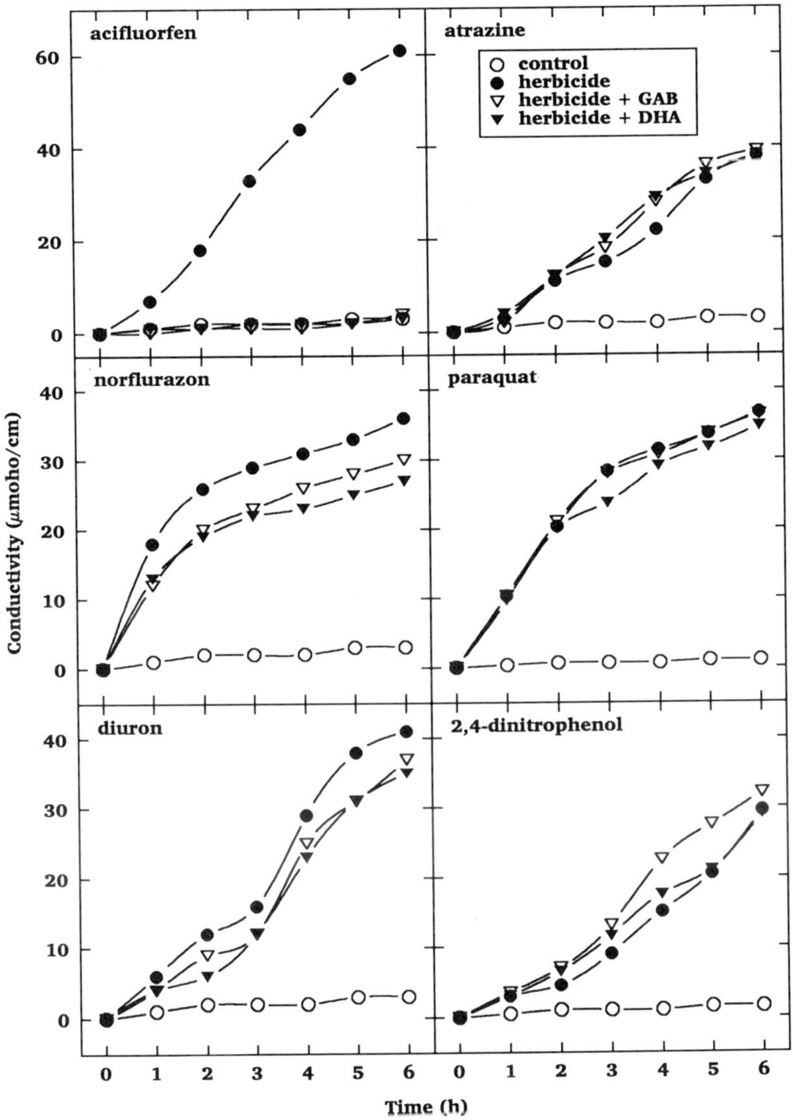

Figure 2. Effects of six herbicides on electrolyte leakage from cucumber cotyledon discs and their interactions with two porphyrin synthesis inhibitors, gabaculine (GAB) and 2,6-dioxoheptanoic acid (DHA). Tissues were incubated for 20 hr in darkness in 0.1 m*M* herbicide (acifluorfen was at 50 μ*M*) with or without 1 m*M* porphyrin synthesis inhibitor. Time zero is the time of initial exposure to continuous light (275 μmol/m^2/sec) after dark incubation. Acifluorfen,5-[2-chloro-4-(trifluoromethyl)phenoxy]-2-nitrobenzoic acid (acifluorfen-methyl is the methyl ester of acifluorfen); diuron,3-(3,4-dichlorophenyl)-1,1-dimethylurea; atrazine,2-chloro-4-(ethylamino)-6-(isopropylamino)-*s*-triazine; norflurazon,4-chloro-5-(methylamino)-2-[3-(trifluoromethyl)phenyl]-3(2*H*)-pyridazinone; oxadiazon,3-[2,4-dichloro-5-(1-methyloxy)phenyl]-5-(1,1-dimethylethyl)-1,3,4-oxadiazol-2(3*H*)one; paraquat,1,1′-dimethyl-4,4′-bipyridylium salt.

The conductivity probe should be calibrated periodically with solutions of known electrical conductance.

The stronger the electrical conductance of the test solution, the more difficult it will be to measure small changes in leakage of electrolytes. In such cases, measurement of leakage of a specific compound from the cells or tissues should be considered.

REFERENCES

Duke, S. O., Lydon, J., and Paul, R. N., Oxadiazon activity is similar to that of *p*-nitro-diphenyl ether herbicides. *Weed Sci.*, 37, 152, 1989.

Matsumoto, H. and Duke, S. O., Acifluorfen-methyl effects on porphyrin synthesis in intact *Lemna pausicostata* Hegelm. 6746 plants. *J. Agric. Food Chem.*, 38, 2066, 1990.

Nandihalli, U. B., Leibl, R. A., and Rebeiz, C. A., Photodynamic herbicides. VIII. Mandatory requirement of light for the induction of protoporphyrin IX accumulation in acifluorfen-treated cucumber. *Pestic. Sci.*, 31, 9, 1991.

Kenvon, W. H., Duke, S. O., and Vaughn, K. C., Sequence of herbicidal effects of acifluorfen on ultrastructure and physiology of cucumber cotyledon discs. *Pestic. Biochem. Physiol.*, 24, 240, 1985.

Sato, R., Nagano, E., Oshio, H., and Kamoshita, K., Activities of the *N*-phenyl imide S-23142 in carotenoid-deficient seedlings of rice and cucumber. *Pestic. Biochem. Physiol.*, 31, 213, 1988.

Photosynthetic Electron Transport

CHAPTER 10

A Simple Fluorometric Detection of Photosystem II Inhibitors

Hermann M. Gleiter and Gernot Renger

INTRODUCTION

Many commercially available herbicides inhibit plant growth by inhibition of photosynthetic activity at the level of photosystem II (PSII). Most of the PSII herbicides, comprising triazine, urea, and phenolic type herbicides, inhibit the electron flow on the acceptor side of photosystem II by binding to the Q_B pocket, thereby blocking the reoxidation of Q_A^- by the plastoquinone pool (Renger, 1986).

The chorophyll fluorescence can provide a useful tool for screening of PSII herbicides because (1) it is easily detectable in whole leaves (noninvasive monitoring) and (2) the quantum yield ϕ_F depends on the functional state of photosystem II. In respect to PSII herbicides, the relation between ϕ_F and the redox state of Q_A is of special interest (ϕ_F is low at Q_A and high at Q_A^-). Upon illumination of dark-adapted samples, Q_A becomes reduced and ϕ_F increases. The light-induced ϕ_F rise is markedly faster if Q_A^- reoxidation is blocked by PSII herbicides. Based on this difference a simple method is described for screening of PSII herbicides. The principles are outlined for the simplified system of spinach thylakoids.

0-87371-539-X/93/$0.00 + $.50

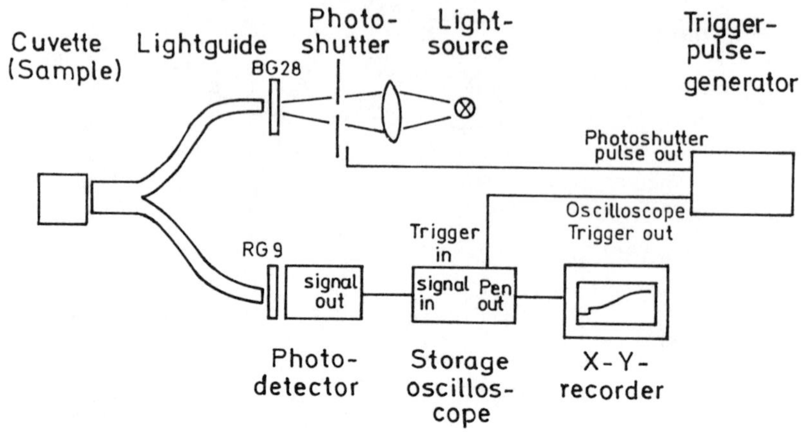

Figure 1. Experimental setup for recording of the time course of fluorescence quantum yield changes (for details see text).

PROCEDURE

Materials

Thylakoids were prepared from local market spinach according to Winget et al. (1965) and stored at −80°C. Rethawed samples were resuspended in pH 7.5 buffer (20 m*M* HEPES (4-(2-hydroxyethyl)-1-piperazine ethane sulfonic acid)/NaOH, 10 m*M* NaCl), and the I_{50} values for diuron (3-(3,4-dichlorophenyl)-1,1-dimethylurea) and atrazine (2-chloro-4-(ethylamino)-6-(isopropylamino)-*s*-triazine) were determined as described below.

Experimental Setup

Fluorescence induction curves were recorded with a setup described in Figure 1. The actinic light beam from a halogen light source (100 W) passing through a blue filter (BG28 from Schott, D-6500 Mainz, Germany; λ_t <520 nm) was switched on and off by an electronic photoshutter (2 ms rise time) and focused on one arm of a two-armed light guide. Actinic light intensity was adjusted by a combination of neutral filters (NG from Schott). The other end of the light guide was fixed to the bottom of the cuvette containing the sample. The resulting red fluorescence emission guided via the second arm of the light guide was detected either by a photodiode or a photomultiplier. The detection system was shielded from blue actinic stray light by a red cutoff filter (RG9 from Schott; λ_t > 710 nm). This ensures a complete separation of the actinic light from the fluorescence emission. Furthermore, the cuvette was carefully protected from any incidental irradiation besides the actinic light. The signal from the photodetector was recorded in a storage oscilloscope (band width > 1 kHz) and plotted on an x-y recorder for documentation and evaluation.

Measurement Protocol

The cuvette with a sample volume of 2 mL was filled with suspensions of dark-adapted ($t_D > 30$ min) thylakoids (final concentration 10 μM chlorophyll) and aliquots of herbicide stock solutions giving rise to the desired concentration (final methanol concentration in the cuvette <3% w/v) and were kept 5 min in complete darkness before the photoshutter was opened for recording the fluorescence induction. The time domain of monitoring was usually 15 sec.

THEORY

Although the fluorescence level in photosystem II of higher plants is determined by many parameters (Renger and Schreiber, 1986), valuable information on the action of PSII herbicides can be obtained on the basis of a rather simple method. This will be exemplified for thylakoid suspensions. Under the conventional experimental conditions (actinic light intensity) used for monitoring the fluorescence induction curves, the quantum yield is almost entirely dependent on the redox state of Q_A. In this case, differences in the rise kinetics of the fluorescence induction curve can be used to determine the portion of inhibited photosystem II units at a given herbicide concentration (for a theoretical treatment of induction curves see Renger and Schulze, 1985).

Figure 2 depicts a schematic representation of the fluorescence induction curve of control spinach thylakoids and of samples which are partially or completely inhibited by the herbicide diuron. The most prominent feature of these curves is the much steeper rise in inhibited thylakoids compared with the control. Therefore, depending on the experimental conditions (i.e., the

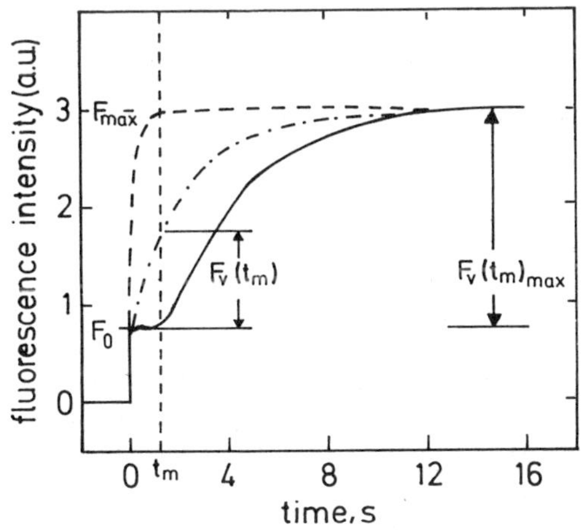

Figure 2. Schematic depiction of fluorescence induction curves in isolated spinach thylakoids with completely inhibited (- - - -), partially inhibited (- · · · -), and noninhibited (——) (control) electron transport.

actinic light intensity, etc.), a characteristic time t_m can be defined where the induction curve of completely inhibited samples is already near its maximum level F_{max} (maximum fluorescence, see Figure 2, for abbreviations) while the fluorescence of intact thylakoids is still close to its initial level F_0. Accordingly, the fluorescence emission at t_m is low in noninhibited PSII units and high in units blocked by herbicides.

If for the sake of simplicity all PSII are assumed to be isolated units, and if we neglect the actual deviations from F_0 and F_{max}, the total fluorescence at t_m for each induction curve is the sum of the fluorescence emission from inhibited and noninhibited units:

$$F(t_m) = pF_{max} + (1 - p)F_0 \quad (1)$$

where p is defined as the fraction of inhibited units.

$$\text{With} \quad F_v(t_m) = F(t_m) - F_0 \quad (2)$$

It follows that

$$p = ((F(t_m) - F_0)/(F_{max} - F_0) = F_v(t_m)/(F_{max} - F_0) \quad (3)$$

In completely inhibited thylakoids ($p = 1$) Equation 3 gives

$$F_v(t_m)_{max} = F_{max} - F_0 \quad (4)$$

In real cases the deviations from F_0 and F_{max} of control and completely inhibited samples, respectively, lead to the relation

$$F_v(t_m)_{max} = r \cdot (F_{max} - F_0) \quad (5)$$

with $r < 1$. These deviations, however, do not affect the calculated values of p if the degree of inhibition is expressed by the equation

$$p = F_v(t_m)/F_v(t_m)_{max} \quad (6)$$

Equation 6 gives a simple formula to calculate the fraction of inhibited PSII units at a given herbicide concentration. It should be mentioned that this simple equation has to be corrected for the energy transfer between PSII units and for additional fluorescence quenching, which might become important with certain herbicides. The first modification leads to a slightly different nonlinear relationship between p and $F_v(t_m)$. In the second case, $F_v(t_m)$ has to be normalized to the actual value of $F_{max} - F_0$ at each herbicide concentration. However, even without these corrections, Equation 6 still provides a good measure for the relative activities of different herbicides, as will be outlined below.

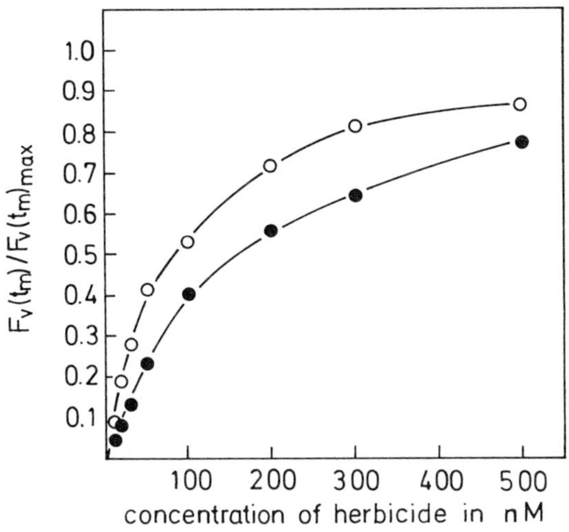

Figure 3. Degree of inhibition as a function of diuron (filled symbols) and atrazine (open symbols) concentration in isolated thylakoids.

RESULTS

Figure 3 shows the dependency of $F_v(t_m)$ on the concentration of the PSII herbicides diuron and atrazine. The p values obtained from the experimental data by using Equation 6 are plotted against the herbicide concentration, neglecting the nonlinear relationship between the portion p of inhibited reaction centers and the variable fluorescence $F_v(t_m)$. The values obtained for the inhibitor concentration at $p = 0.5$ satisfactorily correspond with the I_{50} values gathered from direct electron transport measurements. Therefore, the principles of the method are justified.

COMMENTS

The method described above is based on the differences in the fluorescence induction curves of intact and blocked photosystem II units. It provides a valuable tool for screening of herbicides acting on the PSII acceptor side by blocking the Q_A^- reoxidation. Despite of the rather simplified underlying model for data processing, the method exemplified here for isolated thylakoids was recently shown to provide invaluable information on the interaction of PSII herbicides in leaves (Voss et al., 1984). The method can inter alia be used for (1) monitoring penetrating kinetics of PSII herbicides into leaves, (2) revealing internal metabolic detoxification of PSII herbicides, and (3) screening of unknown compounds for their action as PSII herbicides in leaves.

REFERENCES

Voss, M., Renger, G., Kötter, C. and Gräber, P., Fluorometric detection of photosystem II herbicide penetration and detoxification in whole leaves. *Weed Sci.*, 32, 675, 1984.

Winget, G. D., Izawa, S., and Good, N. E., Stoichiometry of photophosphorylation. *Biochem. Biophys. Res. Commun.*, 21, 438, 1965.

Joliot, P. and Joliot, A., Étude cinétique de la réaction photochimique libérant l'oxygène au cours de la photosynthèse. *C.R. Hebd. Seances Acad. Sci.*, 258, 4622, 1964.

Renger, G. and Schulze, A., Quantitative analysis of fluorescence induction curves in isolated chloroplasts. *Photobiochem. Photobiophys.*, 9, 79, 1985.

Renger, G., Herbicide interaction with photosystem II: recent developments. *Physiol. Vég.*, 24, 509, 1986.

Renger, G. and Schreiber, U., Practical applications of fluorometric methods to algae and higher plant research, in *Light Emission by Plants and Bacteria*, Govindjee, Amest, J., and Fork, D. C., Eds., Academic Press, New York, 1986, p. 587.

CHAPTER 11

In Vivo Measurements of Cytochrome *f* and the P515 Response*

Stephen O. Duke

INTRODUCTION

In vivo methods offer the advantage of allowing for immediate measurement of a compound or process in a plant tissue without the trouble of extraction or sample preparation, during which constituents can be altered to give artifactual results. In this chapter, two *in vivo* spectrophotometric assays are described. Although the methodologies of the two assays are quite similar, the physiological processes that each measures are different.

The first assay is that of the *in vivo* measurement of the oxidation/reduction of cytochrome *f* of chloroplast electron transport. The oxidized form of cytochrome *f* has a lower absorbance at 554 nm than the reduced form, whereas both oxidized and reduced forms have equal absorption at 560 nm (an isosbestic point). Thus, cytochrome *f* can be measured *in vivo* by cycling between the oxidized and reduced forms while measuring the difference in absorption between 554 and 560 nm. This method has been used to detect the effects of paraquat, atrazine (Vaughn and Duke, 1983), acifluorfen (Kenyon et al.,

* This chapter was completed under the auspices of the U.S. Government and is therefore in the public domain.

1985), tentoxin (Duke et al., 1982), and clomazone* (Duke and Kenyon, 1986) on photosynthetic electron transport.

The second method is that of measurement of the P515 response. The 515-nm absorbance change is thought to be due partly to rapid (<0.5 msec) changes in absorbance of both carotenoids and chlorophyll b as a result of charge separation between photosystems (PS) I and II. A much slower (~0.5 sec) absorbance change at 518 nm is thought to be due to acidification of the inner thylakoid membranes by protons liberated by the splitting of water by PSII. As the protons are dissipated (e.g., by darkness), the absorption increase is lost. Under normal conditions, some protons are lost from the thylakoid lumen in a highly controlled fashion through coupling factor complexes. ATP is generated by coupling factor 1 ATPase in the process. Compounds that act as protonophores will result in the reduction of the P515 response, whereas those that block coupling factor ATPase activity prevent proton leakage and enhance the P515 response. Both spectrophotometric methods have been reviewed recently (Duke et al., 1991).

PROCEDURES

Plant Material

These methods can be adapted to any type of plant material. However, leaf material with a width sufficient to cover the sample holder aperture (see below) is most easily utilized. In the examples provided here, several types of plant tissues are used: 7-day-old cucumber cotyledons from which 4-mm discs were cut with a cork borer; mung bean (*Vigna radiata*) primary leaves, as described by Duke et al. (1982); cowpea (*V. unguiculata* L.), as described by Duke and Kenyon (1986); and pea leaf tissue, as described by Vaughn and Duke (1985). The cucumber cotyledon discs were washed in 1% sucrose, 1 m*M* 2-(morpholino)-methanesulfonic acid buffer (pH 6.5), or 0.1% sucrose (no buffer) and then floated in 6-cm petri dishes on 5 mL of the wash solution with or without the phytotoxin.

Chemicals

Technical grade (>95% purity) acifluorfen and clomazone were gifts of Rohm and Haas Co., Spring House, Pennsylvania, and FMC Corporation, Princeton, New Jersey, respectively. Tentoxin, diuron (DCMU), and paraquat were purchased from Sigma Chemicals (St. Louis, Missouri).

* 2-[(2-chlorophenyl)methyl]-4,4-dimethyl-3-isoxazolidinone; for the chemical names of other herbicides see Chapter 9.

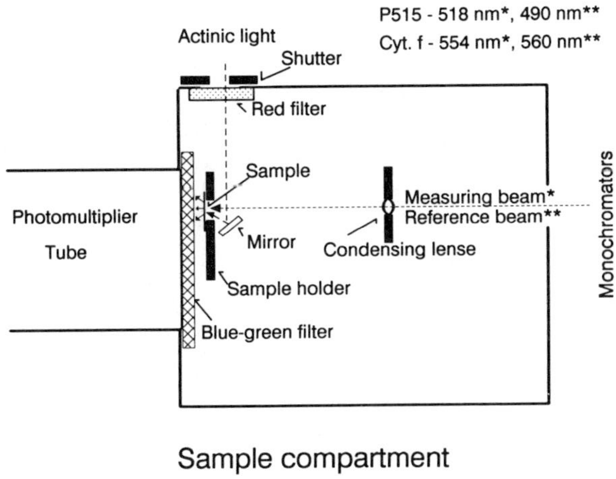

Figure 1. Diagram of spectrophotometer sample compartment for *in vivo* cytochrome *f* oxidation/reduction and P515 response assays.

Equipment

A dual-wavelength spectrophotometer is required for these two assays. A dual-wavelength spectrophotometer has two monochromators, each capable of generating selected wavelengths. The light beams from both monochromators pass through the sample; one is a measuring beam and the other is a reference beam. The reference beam can be chosen at a wavelength that should not change during the assay (an isosbestic point), in order to minimize absorbance drift.

The spectrophotometer should have a sufficiently high linear absorbance range for the absorption spectrum of the plant tissue used to be on scale. The geometry of the photomultiplier tube (PMT) with respect to the sample should be such that the PMT interception of measuring and reference beam light transmitted by the sample is maximized (Figure 1). The PMT should be protected from actinic light by appropriate filters that allow passage of the two spectrophotometric beams, but block wavelengths of actinic light. In the studies* described below, an Aminco® DW-2 or SLM/Aminco® DW-2000 (SLM Instruments, Urbana, Illinois) spectrophotometer was used. A blue-green Corning® glass (No. 9782) filter was used to block the PMT in both the cytochrome *f* and P515 studies.

Actinic light of a wavelength or wavelengths blocked by the PMT filter can be provided by a laser or by a filtered white light source. In the example described here, red actinic light was provided by filtering white actinic light

* Mention of a proprietary product, trademark, or vendor does not constitute a guarantee, warranty, or endorsement by the U.S. Department of Agriculture and does not imply its approval to the exclusion of other products or vendors that also may be suitable.

from an Aminco® side illumination attachment with a Baird-Atomic N-700® sharp cut-on color-glass filter.

The spectrophotometric beams were focused on the sample with a condensing lens. Flat-black, anodized aluminum sample holders with apertures in the holders, which registered with the spectrophotometric beams, were utilized. Samples were mounted on anodized aluminum sample holders with translucent, colorless tape so that the tissue intercepted all spectrophotometric light that passed through the tissue holder.

Procedure

A spectrophotometer slit width of 2 nm was used in all measurements described here. A wider slit width reduces resolution, and a narrower slit width increases noise because of beam attenuation. The narrowest slit width that will allow measurement without unacceptable noise should be used. Measuring beams were 554 and 518 nm and the reference beams were 560 and 490 nm for cytochrome *f* and P515 measurements, respectively.

After treatment of the plant tissues with the phytotoxin, the tissue is taped to the tissue sample holder, ensuring that the tissue completely occludes the beam aperture. This is extremely important in that any light leakage around the sample will result in an attenuated signal. The tissue should be dry because rapid moisture loss during measurement will cause signal drift and noise. The holder is then placed into the spectrophotometer, and the measuring and reference beams are balanced to a point within the selected absorbance range. The measurement is then begun with a baseline, after which actinic light is turned on and then off (preferably with a shutter) at a suitable time intervals to obtain repeatable absorbance changes. Measurements should be made without a sample or with a piece of white paper as a sample in order to be certain that no actinic light is reaching the PMT or that turning the actinic light on and off has no electrical effect on the PMT signal.

RESULTS

Cytochrome *f* Oxidation/Reduction

Figure 2A shows the effect of paraquat and DCMU on the oxidation/reduction of cytochrome *f*. The amplitude of the signal is not significantly affected by either compound. However, the rate of dark reduction is increased by paraquat and reduced by DCMU. The DCMU effect is due to blockage of electron flow to cytochrome *f* from photosystem II. The mechanism of the paraquat effect is not as certain. Paraquat may directly reduce cytochrome *f*.

Extensive damage to the chloroplast will result in loss of the cytochrome *f* oxidation/reduction signal, allowing *in vivo* monitoring of chloroplast dysfunction. For example, acifluorfen causes rapid loss of this signal after acifluorfen-treated tissues are exposed to light and extensive peroxidative damage occurs (Figure 3A).

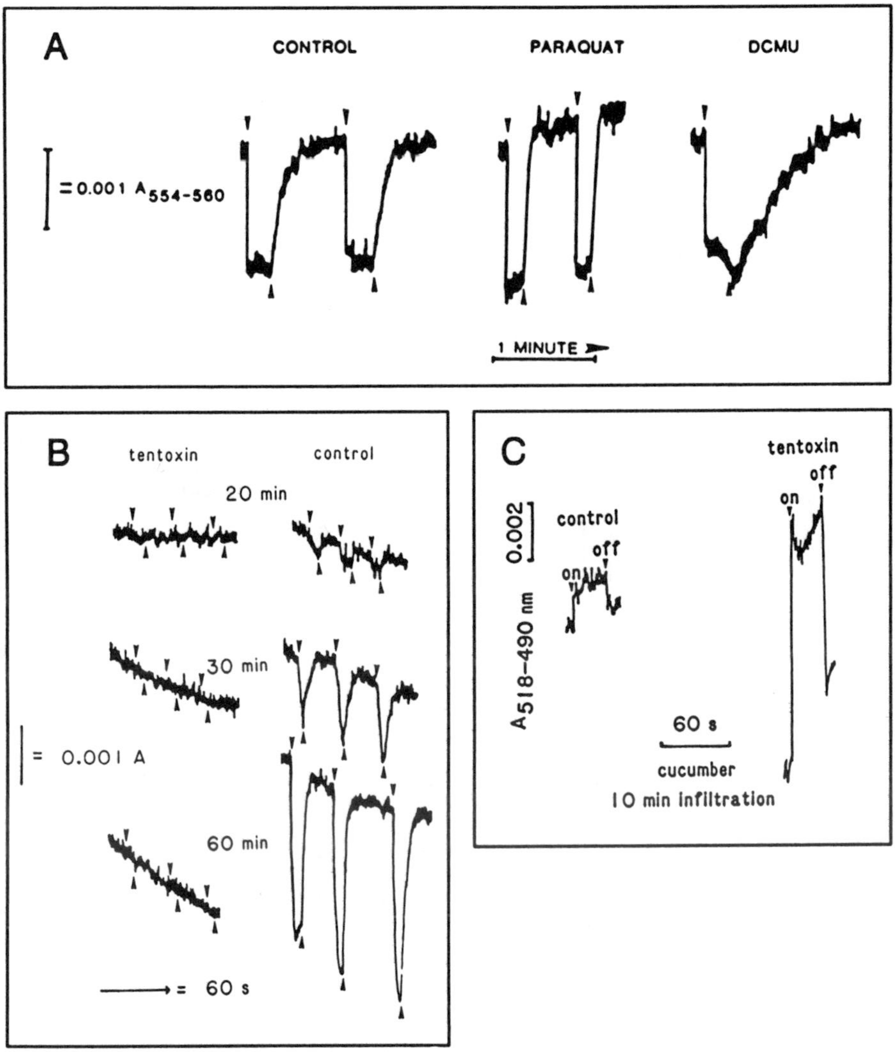

Figure 2. Effects of various phytotoxins on *in vivo* cytochrome *f* oxidation/reduction and on the P515 response. (A) Effects of paraquat and DCMU (diuron) on the *in vivo* cytochrome *f* oxidation/reduction signal in pea leaf discs. Arrows down indicate actinic light on and arrows up indicate light off. (Redrawn from Vaughn, K. C. and Duke, S. O., *Plant Cell Environ.*, 6, 13, 1983. With permission.) (B) Effects of tentoxin on the development of the *in vivo* cytochrome *f* oxidation/reduction signal during greening of mung bean primary leaves. Measurements were made 20, 30, and 60 min after exposure of dark-grown tissue to continuous light. Upper arrows indicate light on and lower arrows indicate light off during measurement. Tentoxin treatment was achieved by imbibing the seeds in 80 μM tentoxin for 24 hr. (Redrawn from Duke, S. O., Wickliff, J. L., Vaughn, K. C., and Paul, R. N., *Physiol. Plant.*, 56, 387, 1982. With permission.) (C) Effect of 80 μM tentoxin on the P515 electrochromic transient in cucumber cotyledon tissue. Actinic red light was turned on and off, as indicated.

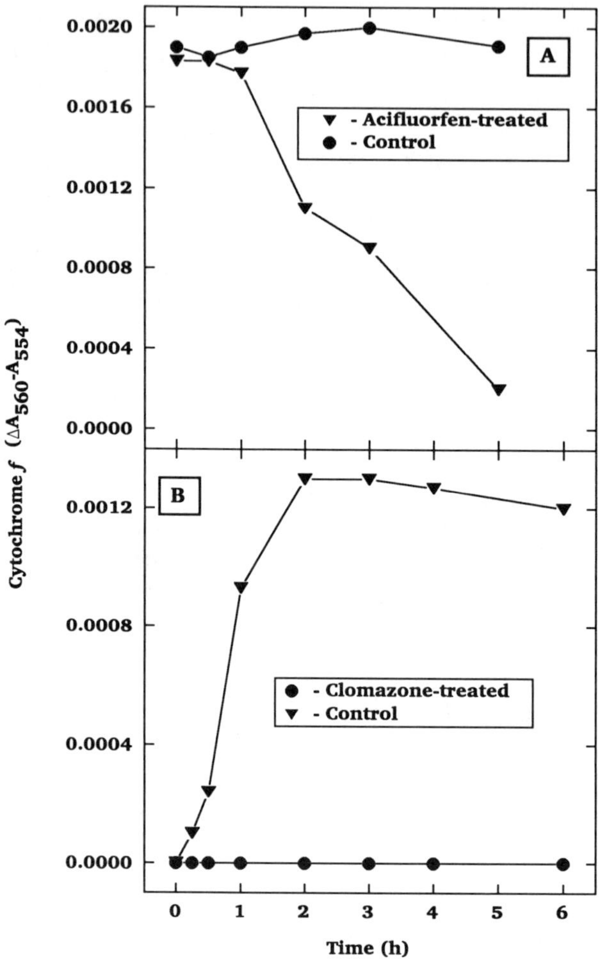

Figure 3. Effects of two herbicides on *in vivo* cytochrome *f* oxidation/reduction. (A) Changes in the *in vivo* cytochrome *f* oxidation/reduction signal in acifluorfen-treated and control cucumber cotyledon discs during a 5-hr illumination period. Tissues were floated on buffer or buffer with 30 μM acifluorfen for 20 hr in darkness before exposure to light (420 μmol/m^2/sec white light). (Redrawn from Kenyon, W. H., Duke, S. O., and Vaughn, K. C., *Pestic. Biochem. Physiol.,* 24, 240, 1985. With permission.) (B) Development of cytochrome *f* oxidation/reduction capability in primary leaves of control and chlomazone-treated, dark-grown, 6-day-old cowpea (*Vigna unguiculata* L.) seedlings during 24 hr of greening. Clomazone treatment was achieved by imbibing the seeds in 0.5 m*M* clomazone for 24 hr before planting. (Redrawn from Duke, S. O. and Kenyon, W. H., *Pestic. Biochem. Physiol.,* 25, 11, 1986. With permission.)

Conversely, chloroplast development from an etioplast results in the appearance of this signal, and any phytotoxin that interferes with greening or chloroplast development will interfere with its appearance. For example, tentoxin, a fungal phytotoxin that prevents uptake of certain nuclear-encoded proteins by plastids and inhibits photophosphorylation, completely prevents the appearance of this signal (Figure 2B). Similarly, clomazone, a carotenoid synthesis inhibitor, can also prevent or delay the appearance of the signal (Figure 3B). In both of these cases it should be noted that the appearance of the cytochrome *f* oxidation/reduction signal is extremely rapid, appearing within minutes of exposure of dark-grown tissues to light.

P515 Response

No commercial herbicides are known to directly affect the P515 response. However, this assay has not been utilized to study herbicides. Studies in this laboratory have shown that tentoxin, a CF_1 ATPase energy transfer inhibitor, profoundly increases the amplitude of this measurement (Figure 2C). As this assay has been done in this laboratory, the two components of the response cannot be separated. From what is known of this response and of the mechanism of action of tentoxin, it can be suggested that the observed effect is due to blockage of dissipation of the proton gradient through inhibition of CF_1 ATPase. Thus, the effect is on the slow portion of the response. Other compounds known to affect thylakoid proton gradients also affect this measurement.

COMMENTS

After the instrumentation is in place, these methods are quick and simple. However, access to a rather sophisticated spectrophotometer in an unusual configuration is required. Another limitation is that each measurement represents only one sample of one plant or plant part. Therefore, many measurements may have to be made and averaged in order to determine subtle differences. Furthermore, the measurements can only be used for comparative purposes. The exact nature of the chromophore(s) of the P515 effect is not known; even if it was, *in vivo* spectrophotometric measurements cannot accurately quantitate pigments. Nevertheless, these assays can be valuable assays of two *in vivo* processes. They can be utilized to examine the effects of phytotoxins on mature chloroplasts and on the development of chloroplasts.

REFERENCES

Vaughn, K. C. and Duke. S. O., *In situ* localization of the sites of paraquat action. *Plant Cell Environ.*, 6, 13, 1983.

Kenyon, W. H., Duke, S. O., and Vaughn, K. C., Sequence of herbicidal effects of acifluorfen on ultrastructure and physiology of cucumber cotyledon discs. *Pestic. Biochem. Physiol.*, 24, 240, 1985.

Duke, S. O., Wickliff, J. L., Vaughn, K. C., and Paul, R. N., Tentoxin does not cause chlorosis in greening mung bean leaves by inhibiting photophosphorylation. *Physiol. Plant*, 56, 387, 1982.

Duke, S. O. and Kenyon, W. H., Effects of dimethazone (FMC 57020) on chloroplast development. II. Pigment synthesis and photosynthetic function in cowpea (*Vigna unguiculata* L.) primary leaves. *Pestic. Biochem. Physiol.*, 25, 11, 1986.

Duke, S. O., Duke, M. V., Sherman, T. D., and Nandihalli, U. B., Spectrophotometric and spectrofluorometric methods in weed science. *Weed Sci.*, 39, 505, 1991.

CHAPTER 12

Inhibition of the Photosynthetic Electron Transport System

Peter Böger

INTRODUCTION

Many herbicides exert their phytotoxic action by interfering with the photosynthetic electron transport. Most of them bind to the D1 protein, a 32-kDa polypeptide belonging to the photosystem (PS) II core complex, which is tightly integrated into the thylakoid. These inhibitors prevent the light-driven reduction of plastoquinone (see Böger and Sandmann, 1990, for review). Another inhibition site, although of minor importance, is the cytochrome b_6/f complex between the two PS (Figure 1). It has been shown that diphenyl ethers can bind at D1 and/or at this complex according to substituents (Draber et al., 1981). Detailed biochemical studies, however, have only been published for the D1 binding domain, particularly with respect to herbicide resistance phenomena and the molecular architecture of D1. Although during the last 30 years many compounds have been produced that inhibit electron transport (and many of them have been developed into herbicides), much interest is still focused on such inhibitors. No herbicides are known which inhibit photosynthesis in its carbon dioxide fixation reactions.

No basic differences in inhibition behavior have been observed within the oxigenic photosynthesis of the plant kingdom. Thylakoids can be easily prepared from photosynthetic cells like microalgae or leaves. The cell-free

0-87371-539-X/93/$0.00 + $.50

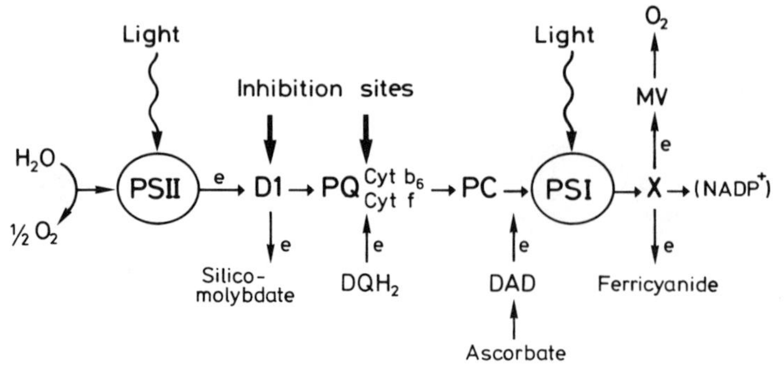

Figure 1. Simplified scheme of the photosynthetic electron transport system with reaction sites for inhibitors, electron donors, and acceptors. Redox carriers of the electron transport system: D1, plastoquinone oxidoreductase, a 32-kDa polypeptide in the core complex of PSII, which binds most of the so-called "photosystem-II herbicides"; PQ, plastoquinone; Cyt b_6, Cyt *f*, cytochromes in the cytochrome b_6-*f* complex; PC, plastocyanin; X, redox components of PSI; PSII, I denote the pigment complexes of the PSII and I absorbing incident light quanta, respecitvely. The water-splitting complex is indicated by the bent arrow at the left. Inhibition of the D1 protein can be relieved by the electron acceptor silicomolybdate. For other abbreviations see Table 1; ferricyanide is potassium hexacyanoferrate(III).

thylakoid system as described here allows for a fast, simple, and reproducible measurement of inhibitory activity of new compounds.

It should be noted that the thylakoid cell-free assay system is the method of choice vs. a cellular one (for instance, the green alga *Scenedesmus*) since it produces information on both inhibitory activity and inhibition sites. A cellular system also includes uptake barriers, although these are rare. The isolated D1 protein — which may be prepared from *Escherichia coli* cloned with the *psb*A gene — cannot be used in an assay since it exhibits binding and inhibition properties only when kept integrated into the PSII core complex.

PROCEDURE

Thylakoids from Higher Plants

Spinach (*Spinacia oleracea,* strain Atlanta or other strains) is the species of choice, but pea leaves are also suitable. Wetted leaves (take the medium-sized leaves) can be stored in plastic bags in the refrigerator for several days before use. After removal of the midrib, 15 g of spinach leaf material are homogenized in 40 mL of a medium containing 0.4 *M* saccharose, 50 m*M* Tricine-NaOH buffer, *N*-tris (hydroxymethyl)methylglycine (pH 8.0), 10 m*M* NaCl, and 5 m*M* $MgCl_2$. Medium and mixer must be cold. Homogenization time is 3 to 4 sec, and any strong mixer can be used (e.g., the Sorvall Omnimixer or a kitchen mixer). The homogenate is pressed through four to

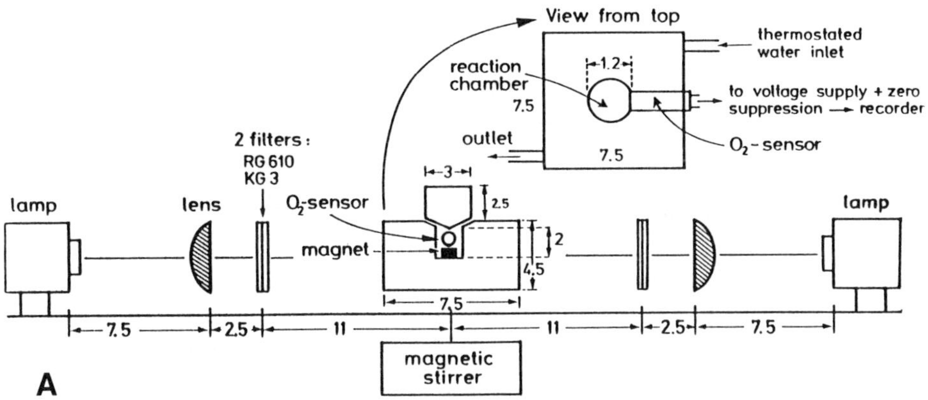

Figure 2A. Setup to measure oxygen produced or consumed in photosynthetic electron transport. Built by the workshop, University of Konstanz, Germany; all figures are given in centimeters. *Lamps:* Model 030123 from Spindler and Hoyer, D-3400 Göttingen, Germany, each equipped with a 12-V halogen bulb. Light is focused into the reaction chamber to give an intensity of about 1000 W/m². *Filters:* On each side a red cut-off filter RG 610 and a heat filter KG 3, all 2 mm thick (from Schott Co., D-6500 Mainz, Germany), are mounted; focal length of the lens is 80 mm, with a 40-mm diameter. *O_2 sensor:* Model 5331 from Yellow Springs Instrument Co., Yellow Springs, OH 45387. *Reaction chamber:* Volume about 2 mL; some of the reaction medium is deliberately removed by the conical stopper to get rid of interfering air bubbles. Furthermore, this type of stopper conveniently allows additions during the assay. After each measurement, which lasts about 2 min, the reaction medium is sucked off by a water pump; the chamber is rinsed with distilled water and is then ready for the next assay. The temperature of the chamber is kept at 20°C by a circulating thermostated water bath as indicated. Everything is made from acryl glass (plexiglass, 7 mm). This instrument is built for quick routine assays. Instruments for oxygen measurements are also commercially available, e.g., from Hansatech Ltd., Norfolk, U.K.

six layers of cheese cloth and is centrifuged for 1 min at 4000 × *g* (5 to 6°C). The pellet is suspended in 1 mL reaction medium containing 0.1 *M* saccharose, 50 m*M* Tricine-NaOH (pH 8.0), and 5 m*M* $MgCl_2$. The reaction is followed with an oxygen electrode, using a 2-mL reaction volume chamber. A useful instrument is built by the University of Konstanz workshop (see Figure 2, also for filters and light intensity).

The assay mixture for higher plant thylakoids includes:

- 1.95 mL reaction medium
- 5 μL 0.4 *M* NH_4Cl
- 30 μL thylakoid suspension, containing approximately 1 mg chlorophyll per milliliter
- 5 to 20 μL inhibitor solution (generally dissolved in methanol or acetone stock solutions of 10^{-2} *M* and less)

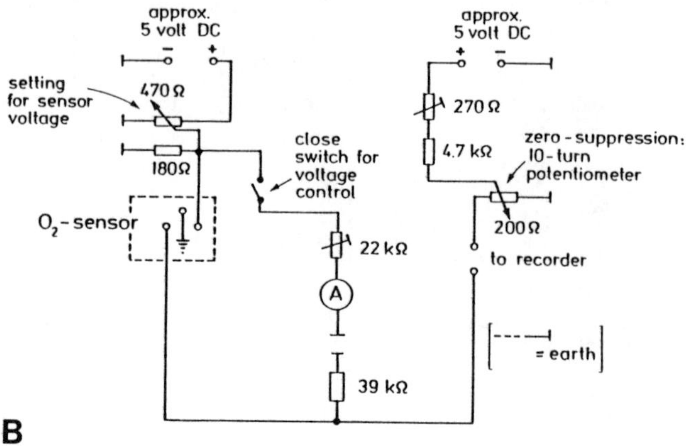

B

Figure 2B. Circuit diagram for the oxygen sensor: voltage supply and zero suppression. About 5 V direct current can be obtained using a mercury battery. For voltage control, the amperemeter (A) should read 60 μA to have a polarizing voltage of about 800 mV at the O_2 sensor.

Thylakoids from Algae

There are only a few eukaryotic species which can be grown easily and will deliver active isolated thylakoids. One of these is *Bumilleriopsis filiformis* Vischer (see Böger, 1971, for growth; a stock culture can be provided by this laboratory). A 200-mL volume of algal suspension (about 90 μg chlorophyll per milliliter of suspension, which equals about 7.5 × 10^6 cells per milliliter) is washed with 10 m*M* tris(hydroxymethyl)aminomethane (Tris)-HCl buffer, pH 8.0, including 15 m*M* NaCl. Homogenization medium (7.5 mL) is added, consisting of 0.35 *M* saccharose, 0.07 *M* Tricine-NaOH (pH 8.0) 3 m*M* Na-pyrophosphate, 1 m*M* $MgCl_2$, 0.5 m*M* EDTA, 3.5% (w/v) polyvinylpyrrolidone, 0.8% bovine serum albumin, 0.4 m*M* ascorbic acid, and 0.04% mercaptoethanol. After mixing with 30 mL glass beads (0.45 to 0.5 mm diameter), the cells are broken in a homogenizer (Braun AG, Melsungen, Germany) for 1 min total homogenization time, with 5-sec intervals under intensive cooling with CO_2 snow. The homogenate, freed from the beads by a suction funnel, is gently centrifuged (2.5 min at 150 × *g*), the supernatant centrifuged again for 5 min at 17,000 × *g*, and the pellet suspended in cold 0.45 *M* saccharose containing 8 m*M* Tricine-NaOH (pH 8.0), 0.4 m*M* Na-pyrophosphate, and 0.1 m*M* $MgCl_2$.

The reaction medium consists of 0.06 *M* saccharose, 50 m*M* 4-(2-hydroxyethyl)-1-piperazine ethane sulfonic acid (HEPES)-NaOH (pH 8.0), 2 m*M* $MgCl_2$, and 1 m*M* Na-K-phosphate. The assay mixture is the same as described for spinach thylakoids in the section on thylakoids from higher plants.

RESULTS

Three assays may be performed to determine possible activities of (new) compounds and their inhibition domain (Figure 1 and Table 1; see these also for chemical names).

1. Inhibition of the D1 protein: Include either 1 m*M* potassium ferricyanide (O_2 is produced) or 0.25 m*M* methylviologen (O_2 is consumed). Using 1 μ*M* diuron, check that the reaction can be completely inhibited. This control is necessary when using the test with methylviologen. Include a control with 5 to 20 μL of solvent without herbicide, although the effect of the solvent is generally small. Run an experiment ±1 m*M* NH_4Cl to check for thylakoid integrity.
2. Inhibition of the cytochrome b_6/f complex: Include 0.5 m*M* durohydroquinone, 0.25 m*M* methylviologen, and 0.5 m*M* KCN. Note that the diphenyl ethers of Table 1 inhibit in assay 1 *and* 2, in contrast to the "normal" photosynthesis inhibitors like diuron and ASD 2287, which are effective only in assay 1. Since the rates in *both* assays are inhibited to the *same* extent, it is concluded that these diphenyl ethers block the cytochrome b_6/f complex. To stop electron flow via D1, 1 μ*M* diuron may be included in the assay.
3. Inhibition of PSI: Include 0.1 m*M* diaminodurol, 1.5 m*M* Na-ascorbate, 0.25 m*M* methylviologen, 0.5 m*M* KCN, and 1 μ*M* diuron. Electron transport proceeds through PSI only. As indicated in Table 1, no inhibition is observed. Indeed, no good inhibitors are available for PSI as yet.

Figure 3 demonstrates a reciprocal plot to linearize electron transport inhibition data. These plots, allowing simple regression equations, have a common intersection with the abscissa, yielding I_{50} values for the chlorophyll concentration used in the assay. When true K_i values are required, an extrapolation to zero chlorophyll has to be made by running several assays with constant inhibitor concentration and different chlorophyll contents. K_i and I_{50} values differ somewhat, but generally I_{50} values are sufficient to characterize the inhibitory activity provided the chlorophyll content of the assay is kept within 30 to 50 μg per assay volume. The illumination time is 2 to 3 min; a linear rate is obligatory. There is very little *dark* O_2 consumption, e.g., in assays 2 and 3 of Table 1.

COMMENTS

1. The *Bumilleriopsis* system has the advantage that the algae can be grown with excellent yield year round. The cultivation setup is like that for *Scenedesmus* (see Chapter 8). The activity (rates) of the material is highly reproducible once cells are harvested at the same cell density.
2. Flowering spinach should not be used. Spinach is not easy to grow in the greenhouse or in the open during summer. This does not hold for peas, which can be cultivated in the greenhouse year round. The leaf biomass

Table 1. Measurement of Two Prominent Inhibition Sites in the Photosynthetic Electron Transport System[a]

Inhibitor	Assay 1		Assay 2		Assay 3	
(*M*)	$H_2O \rightarrow MV$	Inhibition (%)	$DQH_2 \rightarrow MV$	Inhibiton (%)	DAD/Ascorbate → MV	Inhibition (%)
Control	177	—	474	—	430	—
Oxyfluorfen, 5×10^5	83	53	218	53	440	0
DNP-INT, 10^{-5}	46	74	146	69	420	0
Diuron, 10^{-7}	67	72	474	0	410	0
ASD-2287, 5×10^{-6}	55	69	460	3	410	0

Note: DQH_2, reduced duroquinone, tetramethyl-1,4-benzohydroquinone; DAD, 3,6-diaminodurol, 2,3,5,6-tetramethyl-*p*-phenylenediamine; MV, methylviologen or paraquat, 1,1′-dimethyl-4,4′-bipyridylium dichloride; oxyfluorfen, 2-chloro-1(3-ethoxy-4-nitrophenoxy)-4-(trifluoromethyl)benzene; DNP-INT, 2,4-dinitro-1-(2-iodo-3-methyl-4-nitro-6-isopropylphenoxy)benzene; diuron, 3-(3,4-dichlorophenyl)-1,1-dimethylurea; ASD-2287, 2-chloro-4-hydroxy-2,5-diiodopyridine.

[a] Assays 1 and 2 utilize isolated *Bumilleriopsis* thylakoids uncoupled with NH_4Cl, unlike assay 3. Rates are in micromoles O_2 per milligram chlorophyll per hour. Assay 3 refers to photosystem I.

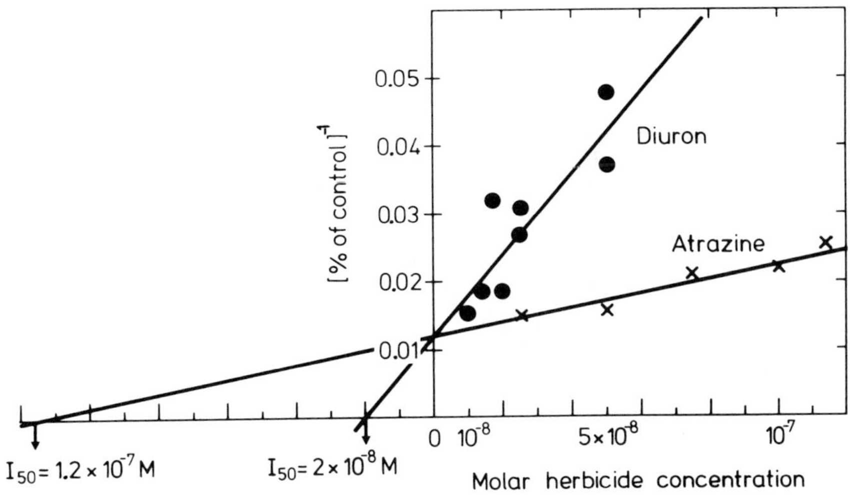

Figure 3. Determination of I_{50} values of inhibitors of photosynthetic electron transport (diuron and atrazine) using isolated thylakoids from the alga *Bumilleriopsis filiformis.* System $H_2O \rightarrow$ ferricyanide, uncoupled with NH_4Cl. Reciprocal plot of rates (percent of control) vs. herbicide concentration.[a]

Inhibitor (*M*)	Rate % of Control	1/Rate
Control	100	—
Diruon		
1×10^{-8}	64	0.0156
1.4×10^{-8}	54	0.0185
2×10^{-8}	54	0.0185
2.5×10^{-8}	37	0.0270
2.5×10^{-8}	32	0.0313
5×10^{-8}	27	0.0370
5×10^{-8}	21	0.0480
Atrazine		
2.5×10^{-8}	66	0.0152
5×10^{-8}	64	0.0156
7.5×10^{-8}	48	0.0208
10^{-7}	46	0.0220
2.5×10^{-7}	39	0.0256

[a] Control rate: 253 μmol O_2/mg chlorophyll × hr. Atrazine is 2-chloro-4-(ethylamino)-6-(isopropylamino)-*s*-triazine; for diuron see Table 1.

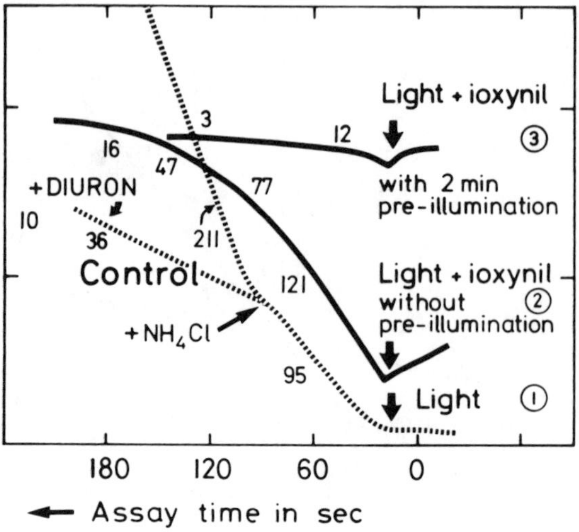

Figure 4. Inhibition of photosynthetic electron flow by "phenolics" herbicides, exemplified with 1 μM ioxynil, 4-hydroxy-3,5-diiodobenzonitrile, and spinach thylakoids (assay system $H_2O \rightarrow$ ferricyanide). Dashed lines indicate the control, without NH_4Cl at start; then NH_4Cl (1 mM) is added as marked by the arrow, and the rate is speeded up. An aliquot is immediately inhibited by diuron (10^{-7} M, tracing 1), while with ioxynil inhibition develops slowly (tracing 2). The final strong steady-state inhibition can be seen immediately after a preillumination of the thylakoids in the reaction mixture before potassium ferricyanide is added (tracing 3). Experiments 2 and 3 include 1 mM NH_4Cl; rates in micromoles O_2 produced per milligram chlorophyll per hour are noted at the tracings.

yield, however, is much less than that of spinach. Spinach can be grown advantageously in a climate chamber at 17–19°C in trays (40 × 10 × 10 cm) using standardized soil (e.g., from Patzer KG, D-6492 Sinntal-Jossa, Germany) with a 12 hr light/12 hr dark regime to prevent flower induction. About 20,000 lux are provided by Philips lamps (SGR 140/400 K). Fluorescent lamps are not recommended.

3. High uncoupled rates, as indicated in assays 2 and 3 of Table 1 and in Figures 2 and 3, are necessary for reproducible I_{50} values. The coupled rate (minus NH_4Cl) is substantially lower. If no rate increase by a factor of about two or higher is obtained after uncoupling with NH_4Cl, the thylakoids should not be used.
4. Phenolics herbicides, like ioxynil or ASD-2287, exhibit low initial inhibition after they are added to the assay (Figure 4). This may lead to erroneous results. Preillumination or longer assay time increases the inhibitory activity. Most probably, this effect is due to impaired access of the ionized form of these compounds to the D1 protein at pH 8 (Mets and Thiel, 1989).

ACKNOWLEDGMENT

This study has been supported by the Fonds des Chemischen Industrie.

REFERENCES

Böger, P., Algal preparations with photophosphorylation activity, *Methods Enzymol.*, 23A, 242, 1971.

Böger, P. and Sandmann, G., Modern herbicides affecting typical plant processes, in *Chemistry of Plant Protection,* Vol. 6, Bowers, W. S., Ebing, W., Martin, D., and Wegler, R., Eds., Springer-Verlag, Berlin, 1990, p. 173.

Draber, W., Knops, H. J., and Trebst, A., Mode of inhibition of photosynthetic electron transport by substituted diphenylethers. *Z. Naturforsch.*, 36c, 848, 1981.

Mets, L. and Thiel, A., Biochemistry and genetic control of the photosystem II herbicide target site, in *Target Sites of Herbicide Action,* Böger, P. and Sandmann, G., Eds., CRC Press, Boca Raton, FL, 1989, p. 1.

CHAPTER 13

Binding of Herbicides at the D1 Protein of the Photosynthetic Electron Transport System

Peter Böger, Gerhard Herrmann, and Andreas Thiel

INTRODUCTION

Photosynthesis inhibitors bind reversibly to thylakoids in a hyperbolic fashion. In a double-reciprocal plot, this curve yields a straight line. Binding, as can be demonstrated with a ^{14}C-labeled herbicide, can be alleviated by competitive replacement by another photosynthesis inhibitor, again in a hyperbolic manner (Mets and Thiel, 1989; Figures 1 and 2). These findings indicate that there is one peptide in the electron transport system responsible for binding of *one* herbicide molecule. This D1 polypeptide, an integral subunit of the photosystem (PS) II core complex, has also been called the "herbicide binding protein". Once one molecule of inhibitor is bound per D1 molecule its catalytic activity is blocked, since plastoquinone is displaced from this enzyme (which may be called a plastoquinone oxidoreductase). Consequently, electron transfer from the primary acceptor Q_A via plastoquinone to the cytochrome b_6/f complex is inhibited.

The 32-kDa D1 protein has (about) 343 amino acids, and in a stretch of about 65 amino-acid residues (from Phe-211 to Leu-275) are located the possible binding partners for an inhibitor molecule (Trebst, 1991). Many mutants are known with changes in this amino acid region, leading to impaired binding of different herbicides. The most prominent example is the exchange

0-87371-539-X/93/$0.00 + $.50

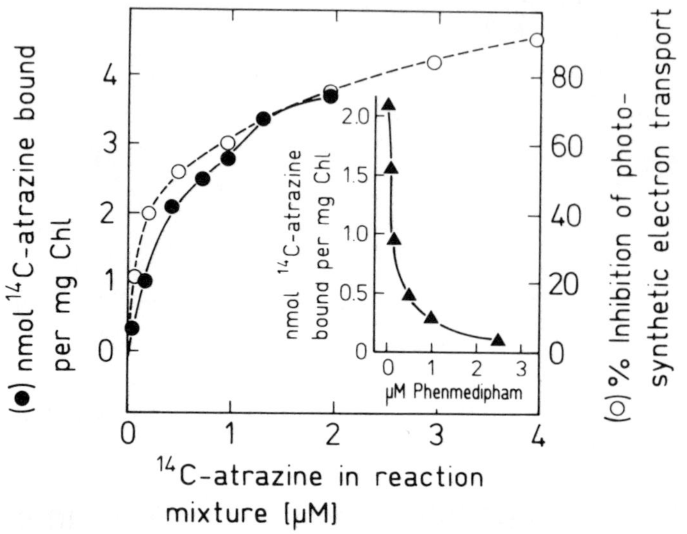

Figure 1. Binding of ^{14}C-labeled atrazine to isolated thylakoids (-●-) with concurrent inhibition of photosynthetic electron transport (H_2O ferricyanide, -○-). Atrazine can be replaced by (unlabeled) phenmedipham, 3-[(methoxycarbonyl)amino]phenyl-(3-methylphenyl)carbamate, or other inhibitors binding at the D1 protein (see inset).

of Ser-264 → Gly-264 leading to a more than 500-fold increase of the binding constant with a concurrent loss of atrazine binding causing resistance against this herbicide. See Thiel and Böger (1984) for a comparative binding study with thylakoids of different resistant mutants.

Several herbicide-resistant mutants have been isolated or produced, also by chemical mutagenesis of the corresponding *psbA* gene, and many laboratories are interested in "transgenic" resistant plants. This contribution presents the methodology to measure (reversible) herbicide binding and to perform (irreversible) photoaffinity labeling of the D1 protein in isolated thylakoids. Both assays depend on ^{14}C-labeled herbicides. It should be noted that binding studies cannot be performed with the isolated (and purified) D1 polypeptide.

Binding studies as described here are of general importance for other herbicides interacting with membrane-bound enzymes — for example, the plastidic phytoene desaturase (carotene biosynthesis) or the protoporphyrinogen oxidase which is instrumental in chlorophyll formation.

PROCEDURE AND RESULTS

Reversible Herbicide Binding

Spinach or *Bumilleriopsis* thylakoids are prepared as described in Chapter 12. The reaction mixture contains in a final volume of 1 mL (use Eppendorf

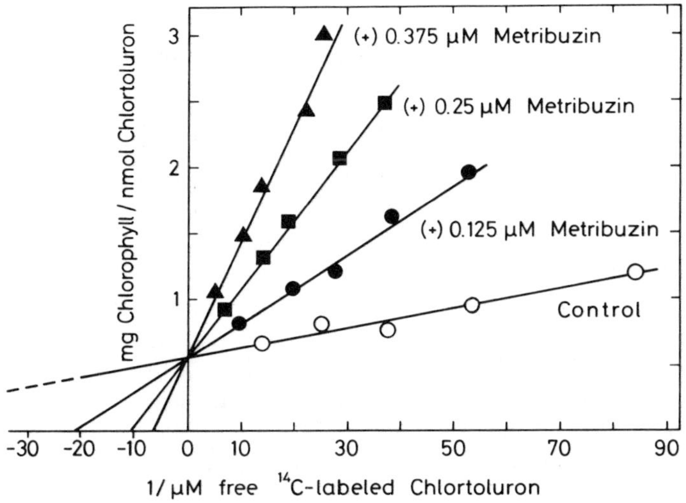

Figure 2. Replacement of the herbicide chlorotoluron, 3-(3-chloro-4-methylphenyl-1, 1-dimethylurea (^{14}C-labeled), by (unlabeled) metribuzin, 4-amino-6-tert-butyl-3-(methylthio)-1,2,4-triazin-5(4*H*)-one, using three different concentrations as indicated. The common intersection of the ordinate, as seen in the double-reciprocal plot,indicates competetive interaction between the herbicides at a common binding site. The maximum number of herbicide molecules bound to this binding site (i.e., the D1 peptide of the photosynthetic electron transport system) is 0.5 mg (= 0.5 µmol) chlorophyll per nanomole chlortoluron = 1 herbicide molecule per 500 molecules of chlorophyll. Note that this figure may vary somewhat due to cultivation conditions and among species. The chlorophyll content of the photosystems depends inversely on light intensity during growth.

centrifugation tubes of 1.5 mL total volume):

- 25 m*M* tris(hydroxymethyl)aminomethane (Tris)-HCl, pH 8.0
- 50 m*M* NaCl
- 5 m*M* $MgCl_2$
- Thylakoids should be equivalent to 70 to 90 µg of chlorophyll
- Each assay sample contains a constant concentration of ^{14}C-labeled herbicide (2 to 5 $\times$ 10^{-2} µ*M*); 5–20 mCi/mmol (= 185–740 MBq) specific activity is appropriate.
- Use five to seven assay samples with increasing concentrations of the corresponding unlabeled herbicide in the range of about 0.1 to 1.0 m*M* depending on binding affinity.

All assays are performed in duplicate. Controls should be run without thylakoids (containing the reaction medium only) and with the solvent (methanol or acetone, which should not exceed 1% v/v). The assays are started by adding the thylakoids to the complete mixture. After vigorous vortexing for 5 sec, the samples are incubated for 10 min and then centrifuged for 5 min at 9000 × *g* (using a small Eppendorf or Beckman centrifuge). All procedures are performed in the cold room under normal illumination. Aliquots of the supernatant (0.9 mL) are mixed with 2.3 mL of water-miscible scintillator. Counts are corrected for quenching, but there is only a little (green) color interfering with the count yield (Thiel and Böger, 1984). A good liquid scintillation counter should reliably measure samples down to 100 dpm per vial.

^{14}C-Labeled atrazine, 2-chloro-4-(ethylamino)-6-(isopropylamino)-*s*-triazine, or diuron, 3-(3,4-dichlorophenyl)-1,1-dimethylurea, including azido analogues (e.g., in atrazine the -Cl is replaced by -N_3, in diuron the 4-Cl by -N_3), generally all ring-labeled, can be purchased from Pathfinder Laboratories Inc., 11542 For Mims Drive, St. Louis, Missouri 63141, from Amersham-Buchler, Gieselweg 1, D-3300 Braunschweig, Germany, or from Sigman Co., D-8024 Deisenhofen, Germany or P.O. Box 14508, St. Louis, Missouri 63178. The companies also offer custom synthesis.

Covalent Photoaffinity Labeling

First, the method allows one to determine the binding protein or (membrane-bound) enzyme of a certain inhibitor; second, the binding region of the enzyme can be determined by chemical degradation of the protein after photoaffinity labeling. A ^{14}C-labeled herbicide molecule has to be modified by introducing an azido group. This molecule is sensitized by ultraviolet (UV) light, N_2 is split off, and the herbicide molecule is covalently bound to an amino acid residue which — presumably — is one of the binding partners of the herbicide or is located in the binding niche of the D1 polypeptide, in close proximity to the binding site of the inhibitor. The thylakoid is subsequently solubilized with sodium dodecyl sulfate (SDS) and the membrane protein separated on SDS gels and visualized by autoradiography (Pfister et al., 1981).

Reaction Mixture

- A 13-mL volume of thylakoid suspension (50 μg chlorophyll (ChL) per milliliter) is placed into a small glass petri dish (6 cm diameter).
- Add azido-modified inhibitor (10 to 50 mCi/mmol) in a final concentration of 0.1 to 1.0 μ*M*; mix thoroughly.
- Irradiate with UV light for 10 min at room temperature with two Sylvania G8T5 germicidal UV lamps (Birchover Instruments Ltd., Letchworth, Herts., U.K.), 10 cm distance from the dishes (366-nm light for azido-atrazine; 254-μm light for azido-diuron).
- Centrifuge for 5 min at 9000 × *g*.
- Suspend the pellet in reaction medium (= first three components of the reaction mixture of the previous section) to give 1 to 1.5 mg ChL per milliliter.
- Mix 1:1 (v/v) with SDS sample buffer (80 m*M* Tris-HCl, pH 6.8; 4% (w/v) SDS; 2% dithiothreitol (DTT); 20% glycerol; 0.02% bromophenol blue).

- Incubate for 30 min at room temperature; then centrifuge aliquots for 5 min at 9000 × *g* in 1.5-mL Eppendorf plastic tubes.

Electrophoresis

- Place 100 μL of the supernatant on an SDS gel, 1.5 mm thick, of the following composition: 17.5% (w/v) polyacrylamide, 0.073% bisacrylamide, 0.38 *M* Tris-HCl (pH 8.7), 0.1% SDS, 0.05% ammonium persulfate, and 10 μL TEMED (*N*′,*N*′-tetramethylethylenediamine per 30 mL.) This is the separating gel, with a length of 12 cm. The spacer (collecting) gel, 1 cm long, contains 5% (w/v) polyacrylamide; 0.13% bisacrylamide, 0.13 *M* Tris-HCl (pH 6.8), 0.1% SDS; 0.05% ammonium persulfate, and 5 μL TEMED per 10 mL.
- Electrophoresis may run overnight at 4°C with a constant current of 0.5 mA/mm². The running buffer consists of 25 m*M* Tris-HCl, (pH 8.4), 190 m*M* glycine, and 0.1% SDS.

Identification

- Shake gels gently in a 0.1% Coomassie Brilliant Blue R-250 solution including 45% methanol and 10% acetic acid (v/v) for 20 to 30 min.
- Decolorize the gel for 2 to 3 hr in 400 mL of a solution containing 45% methanol, 10% acetic acid (v/v), and 1 to 2 Dowex grains, type 1 × 8 (Serva, Heidelberg, Germany). See stained bands in Figure 3 (left).
- Keep the gel for 1 to 2 days at 4°C in a 45% methanol/2.5% glycerol (v/v).
- Shake gel for 20–30 min in the dark in "enhancer" solution for autoradiography (Enlightning from New England Nuclear, Boston, Massachusetts).
- Dry the gel for 2 hr at 80°C on Whatman 3-mm filter paper using a vacuum gel dryer.
- Put an X-ray film on top (Hyperfilm-MP, Amersham, Braunschweig or Kodak-X Omat AR-5) and store in a Kodak-X-Omatic cassette for several days or up to 2 weeks in the deep freeze (−70°C). Process film with Kodak AL-4 fixing solution.
- The photograph is correlated with the Coomassie-stained protein bands of the gel.

COMMENTS

1. Unspecific binding of labeled herbicide may superimpose on the hyperbolic curve of Figure 1. Such curves have the tendency not to reach a plateau, but to steadily increase linearly with higher herbicide concentrations. This linear part, if it is below 20% of total binding, can be safely subtracted from the hyperbolic curve.
2. The advantage of this assay is to measure the free herbicide in the *supernatant* of the section on reversible herbicide binding. The pellet, containing the bound part, can be measured only after a (complicated) burning of the sample in an oxidizer to get rid of the quenching chlorophyll.

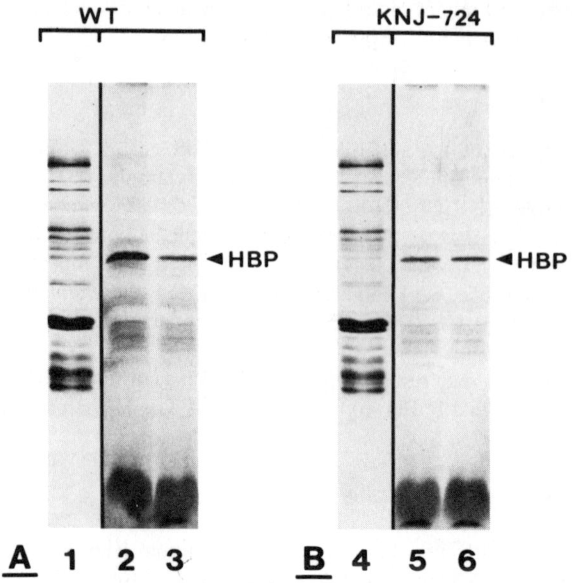

Figure 3. Azido-^{14}C-diuron labeling using isolated thylakoids from (A) the xanthophycean alga *Bumilleriopsis filiformis,* wild type (WT) (see Herrmann et al., 1985) and (B) a mutant KNJ-724. The mutant shows resistance against diuron, exhibiting an I_{50} value about 25-fold higher than that of the wild type. Lanes 1 and 2 demonstrate the Coomassie-stained proteins of the SDS gel; lanes 2 and 5 show the azido-diuron labeling (autoradiography) of the 32-kDa D1 polypeptide. Labeling is less in the mutant (lane 5) than in the wild type (lane 2). The protection experiment with 0.5 μM unlabeled diuron + labeled azido-diuron present exhibits a strong decrease of the label using wild-type thylakoids (lane 3) with little difference between lanes 5 and 6, representing the diuron-resistant mutant.

Accordingly, care must be taken that the supernatants to be counted are colorless. The method is based on the determination of a reproducible difference between the control counts and the counts due to the free herbicide left in the supernatant after binding. Accordingly, this method is only suitable for compounds with a marked binding affinity. Note that the specific activity of the labeled herbicide should not be below 5 mCi/mmol.

3. Unspecific adsorption to the walls of the vessels may cause losses when assaying very active compounds and using low concentrations. Controls are important.
4. Photoaffinity labeling requires higher specific activity than the reversible binding experiments. Requirement of a higher specific activity also depends on a possible increase of the binding constant vs. the mother compound when the azido group has to be attached at an unfavorable position of the herbicide molecule.
5. Protection experiments with unlabeled original herbicides have to be included in photoaffinity assays. Lane 3 of Figure 3 demonstrates that the

intensity of the ^{14}C-azido-diuron-labeled D1 protein band is strongly diminished with cold diuron present, indicating that the same peptide is attacked by both compounds. In the protection experiment unlabeled herbicide is first added to the reaction mixture (see the section on covalent photoaffinity labeling) in the range of 1.0 to 10 μM final concentration (depending on the binding constant); incubation is for 5 min, and then the azido-analogue is added and the experiment continued as described.

ACKNOWLEDGMENT

Our studies were supported by the Fonds der Chemischen Industrie.

REFERENCES

Mets, L. and Thiel, A., Biochemistry and genetic control of the photosystem-II herbicide target site, in *Target Sites of Herbicide Action,* Böger, P. and Sandmann, G., Eds., CRC Press, Boca Raton, FL, 1989, p. 1.

Thiel, A. and Böger, P., Comparative herbicide binding by photosynthetic membranes from resistant mutants. *Pestic. Biochem. Physiol.,* 22, 232, 1984.

Trebst, A., The Molecular basis of resistance of photosystem-II herbicides, in *Herbicide Resistance in Weeds and Crops,* Caseley, J. C., Cussans, G. W., and Atkin, R. K., Eds., Butterworth-Heinemann, Oxford, England, 1991, p. 145.

Pfister, K., Steinback, K. E., Gardner, G. and Arntzen, C. J., Photoaffinity labeling of a herbicide receptor protein in chloroplast membranes. *Proc. Natl. Acad. Sci. U.S.A.,* 78, 981, 1981.

Hermann, G., Thiel, A., and Böger, P., Antiserum against the 33-kDa herbicide-binding protein. *Z. Naturforsch.,* 40c, 814, 1985.

Amino Acid Biosynthesis

CHAPTER 14

Irreversible Inhibition of Glutamine Synthetase from Higher Plants by the Herbicide Phosphinothricin

Remigius Manderscheid

INTRODUCTION

Glutamine synthetase (GS) of higher plants catalyzes the amidation of glutamate:

$$\text{L-Glutamate} + \text{ATP} + \text{NH}_4^+ \rightarrow \text{L-glutamine} + \text{ADP} + \text{P}_i$$

Two isoenzymes have been found in leaves of many plant species (GS_1: located in the cytosol, GS_2: located in the chloroplast) and a third enzyme in the roots (GS_R) (see, e.g., McNally et al., 1983; Manderscheid and Wild, 1986a). The new herbicide phosphinothricin (PPT, also called glufosinate) has been shown to irreversibly inhibit GS by binding at the glutamate site of the enzyme and forming a covalent bond after phosphorylation with ATP (Manderscheid and Wild, 1986b):

$$\text{GS} + \text{PPT} + \text{ATP} \leftrightarrow \text{GS} \cdot \text{PPT} + \text{ATP} \rightarrow \text{GS–PPT–phosphate} + \text{ADP}$$

Studies on the effect of PPT on plant ammonium assimilation can be done by spraying the shoot with the inhibitor and following the decrease of GS

0-87371-539-X/93/$0.00 + $.50

activity in the respective organ. This procedure shows the plant part (or parts) in which GS is actually inhibited. The process at the cell or enzyme level can be studied by the determination of the I_{50} value — i.e., the inhibitor concentration needed for 50% inhibition. As PPT binds to the same site as glutamate, the *in vivo* glutamate concentration, which roughly corresponds to the K_m value of the respective isoenzyme, should be taken into account. The K_m value strongly depends on the isoenzyme, with GS_R and GS_1 having high ($K_m \approx 5$ m*M*) and GS_2 low glutamate affinity ($K_m \approx 15$ m*M*) (Manderscheid and Wild, 1986a).

In the experiment outlined below, the inhibitory action of PPT on GS_R and GS_2 of mustard plants is shown by measuring the decrease of enzyme activity both after treatment of the whole plant and after incubation of the enzyme in root and leaf crude extract with different PPT concentrations.

PROCEDURE

Plant Cultivation

Mustard plants (*Sinapis alba* L.) are grown for 16 to 20 days in expanded clay with Hoagland solution. They are kept at 21°C and about 75% relative humidity and illuminated in a 16:8 hr light-dark regime by fluorescence tubes with photosynthetic active radiation of about 11 $W \times m^{-2}$.

Herbicide

Spraying of the plants is done with the ammonium salt of phosphinothricin ((3-amino-3-carboxy-propyl)-methylphosphinate). Each plant is treated with 2 mL of a solution containing 0.1% detergent and 0.3 mg of PPT, which is equal to about 0.05 kg/ha. For the *in vitro* inhibition study the potassium salt of PPT is used.

Equipment

For GS assay a thermostated water bath and a spectrophotometer are required.

Experimental

Measurement of the decrease of GS activity after PPT application to the plants occurs at several time intervals after treatment; roots and leaves are harvested and immediately assayed for GS activity.

For determination of the I_{50} value, the enzyme assay is run with different PPT concentrations (0.0005 to 0.1 m*M*) and glutamate levels corresponding nearly to the *in vivo* level (root extract: 4 and 6 m*M*; leaf extract: 10 and 20 m*M*). The enzyme activity (percent of control) is plotted against the logarithmic values of the respective PPT concentrations, and the I_{50} is determined from the resulting curve.

Table 1. Time Course of Glutamine Synthetase Activity in Roots and Leaves of Mustard Plants after Phosphinothricin (PPT) Treatment of the Shoot[a]

Time after Application of PPT (hr)	Enzyme Activity							
	Leaves				Roots			
	− PPT		+ PPT		− PPT		+ PPT	
	(a)	(b)	(a)	(b)	(a)	(b)	(a)	(b)
0	408	100%	408	100%	216	100%	216	100%
4	—	—	303	74%	—	—	108	50%
8	—	—	231	57%	—	—	87	40%
24	368	90%	136	33%	224	104%	41	19%
48	350	86%	134	33%	200	93%	37	17%

Note: —, no data.

[a] Enzyme activity is given in (a) μmol/hr × g fresh weight and in (b) % of the initial activity.

For the assay of GS, plant tissue (e.g., mustard, *Sinapis alba*; 0.2 to 0.4 g fresh weight) is ground in a mortar in 4 mL of precooled 50 m*M* tris(hydroxymethyl)aminoethane (Tris)-HCl buffer pH 7.8 containing 0.5 m*M* EDTA, 1 m*M* $MgSO_4$, and dithiothreitol (roots: 5 m*M*; leaves: 10 m*M*). The crude extract is centrifuged 10 min at 20,000 × *g*, and the supernatant is used for GS assay. Enzyme activity is measured at 37°C by the formation of γ-glutamylhydroxamate. The concentrations of the reactants in the 2-mL assay mixture are: 100 m*M* imidazole buffer (pH 7.2), 50 m*M* $MgSO_4$, 10 m*M* EDTA, 10 m*M* NH_2OH, 15 m*M* ATP, 50 m*M* (root extract) or 160 m*M* (leaf extract) glutamate, and 0.1 to 0.2 mL crude extract. The pH of the substrate solution is adjusted beforehand to pH 7 to 8 using KOH. The reaction is started by adding crude extract. After 15 min the assay is stopped with 1.5 mL of an acidic $FeCl_3$ solution (0.37 *M* $FeCl_3$, 0.67 *M* HCl, 0.20 *M* trichloroacetic acid). Fe^{3+} forms a colored complex with γ-glutamylhydroxamate, which is quantified in the supernatant after centrifugation (10 min at 5000 × *g*) by reading the absorbance at 540 nm (1 μmol corresponds to an absorbance change of 0.24). Blanks are stopped immediately after the addition of crude extract. Assay of GS_R and leaf GS, which in mustard only exists as GS_2 (Manderscheid and Wild, 1986a), are linear up to an increase in absorbance of approximately 0.4 and 0.6, respectively.

RESULTS

After spraying the shoots of mustard plants with PPT, a rapid decline of GS activity in both leaves and roots occurs within a few hours (see Table 1). The rate as well as the amount of decrease is greater in the roots than in the leaves. The determination of the I_{50} value at physiological glutamate concentrations yielded only slight differences at the two glutamate levels used for each isoenzyme, but strong differences between GS_2 and GS_R were seen. The I_{50} of the root enzyme ranges between 0.0033 and 0.0042 m*M*, and that of the leaf enzyme ranges between 0.0157 and 0.0167 m*M* (Figure 1).

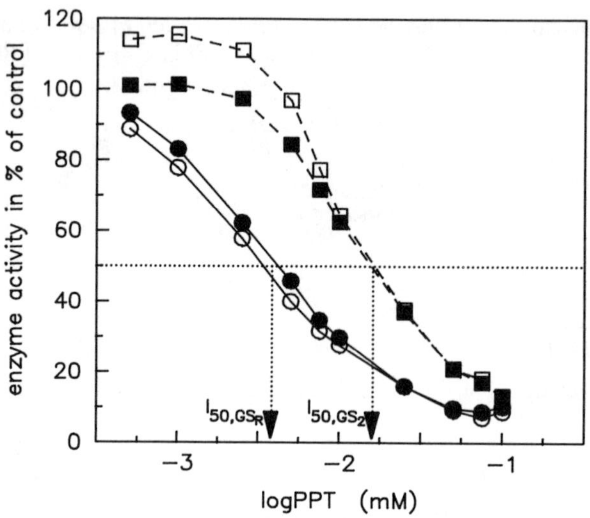

Figure 1. Determination of the I_{50} value by *in vitro* incubation of GS of root (GS_R, circles) and leaf crude extract (GS_2, squares) from mustard with PPT at physiological glutamate concentrations in the enzyme assay. Filled and unfilled symbols indicate the value at the high and low glutamate level in the assay, respectively. Enzyme activity of control (μmol/hr × g fresh weight): root extract, 103 and 140 at 4 and 6 m*M* glutamate, respectively; leaf extract, 110 and 231 at 10 and 20 m*M* glutamate, respectively.

COMMENTS

With the two methods presented, the inhibitory action of PPT on GS of higher plants can be determined rapidly. The assays of GS activity in roots and leaves after PPT application to the shoot demonstrated to what extent the enzyme is affected in both compartments. The amount of inhibition found is connected to the PPT distribution, degradation, etc., and the reactivity of the respective isoenzyme with the inhibitor. The later factor is given by the I_{50} value. Thus, both methods can be used as quantitative biochemical herbicide activity assays. Additionally, their combination can be used to assess whether in the field different effects on the root and leaf enzyme result from isoenzyme properties or other factors. The more sophisticated determination of the reaction rate constant and the K_i values is a large-scale experiment (Manderscheid and Wild, 1986b). Furthermore, it can only be done if the enzyme × inhibitor reaction obeys Michaelis-Menten kinetics. This is always valid for GS_R and usually for GS_2 (e.g., in the wheat plant; Manderscheid and Wild, 1986b). In mustard plants, however, GS_2 shows a positive cooperative binding of glutamate (Manderscheid and Wild, 1986a), which is responsible for the activation of the enzyme at low PPT concentrations seen in Figure 1. Moreover, several C_3 and most of the C_4 plants contain GS_1 and GS_2 in the leaves (McNally et al., 1983), making it unreasonable to determine reaction rate constants in leaf homogenate. These problems can be avoided by the more practical approach, measuring the I_{50} value in plant crude extract at physiological glutamate concentrations, which gives a rough measure of the reactivity between PPT and GS of the respective organ.

REFERENCES

McNally, S. F., Hirel, B., Gadal, P., and Mann, A. F., and Stewart, G. R., Glutamine synthetase of higher plants. Evidence for a specific isoform content related to their possible physiological role and their compartmentation within the leaf. *Plant Physiol.,* 72, 22, 1983.

Manderscheid, R. and Wild, A., Characterization of glutamine synthetase of roots, etiolated cotyledons and green leaves from *Sinapis alba* (L.). *Z. Naturforsch.*, 41c, 712, 1986a.

Manderscheid, R. and Wild, A., Studies on the mechanism of inhibition by phosphinothricin of glutamine synthetase isolated from *Triticum aestivum* L. *J. Plant Physiol.,* 123, 135, 1986b.

CHAPTER 15

The Buckwheat Assay for Glyphosate and Other Inhibitors of Aromatic Biosynthesis

Nikolaus Amrhein and Philippe Roy

INTRODUCTION

Of several classes of herbicides presently known to interfere with essential amino acid biosynthesis, glyphosate (*N*-phosphonomethylglycine), the active component of ROUNDUP® herbicide, was the first shown to have this mode of action. In growth experiments with microorganisms and cultured plant cells, glyphosate toxicity was alleviated by the aromatic amino acids phenylalanine and tyrosine, suggesting interference of the herbicide with the biosynthesis of these aromatic amino acids. The anticipated depletion of the pool sizes of the aromatic amino acids in glyphosate-treated cells or plants was, however, rarely observed. Hypocotyls from etiolated buckwheat seedlings provide a very simple system in which the rapid synthesis of phenylalanine-derived products such as anthocyanin pigments can be induced under suitable experimental conditions. Contrary to the soluble aromatic amino acids in the hypocotyl tissues, the anthocyanins (glycosides of cyanidin) have little, if any, turnover, and their rate of accumulation therefore reflects their rate of synthesis. Furthermore, their extraction and spectroscopic quantitation are remarkably simple. Cyanidin, like other flavonoids, contains two aromatic rings; the A-ring is derived from three acetate units, while the B-ring is derived from a phenylpropanoid unit (Figure 1). The precursor for the latter unit is

0-87371-539-X/93/$0.00 + $.50

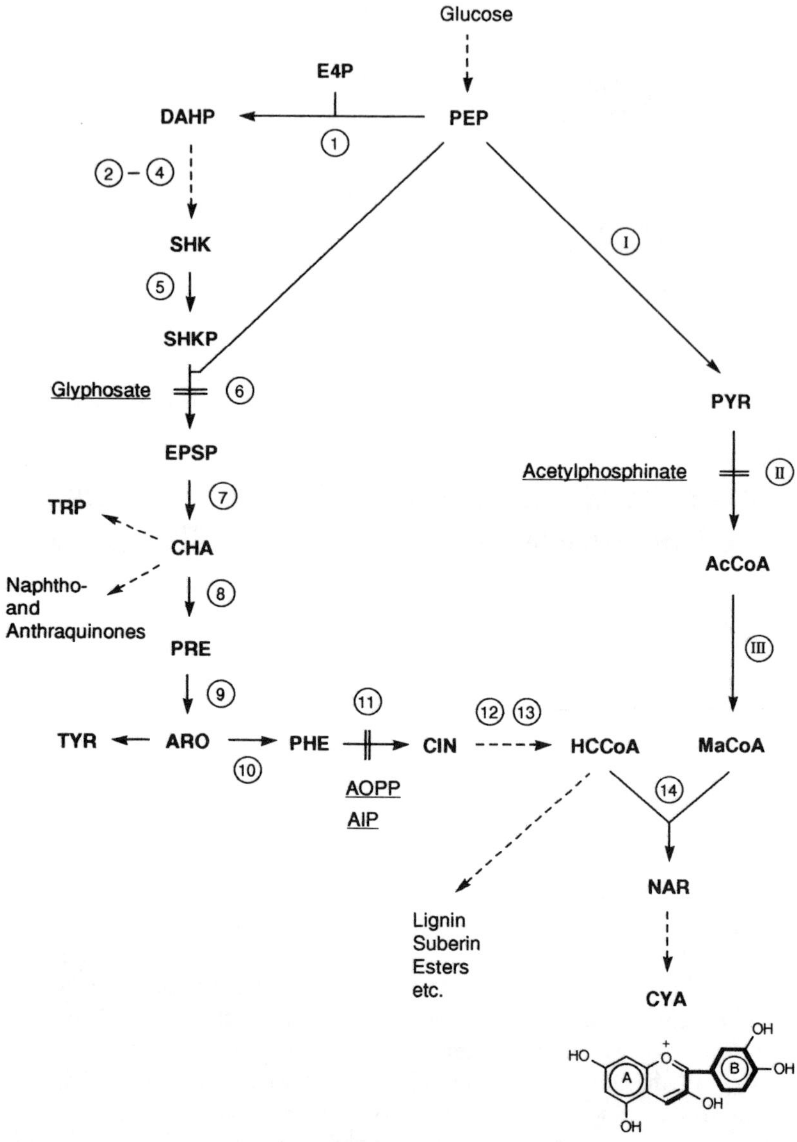

Figure 1. The pathway of cyanidin biosynthesis and sites of action of inhibitors of its synthesis in buckwheat hypocotyls. The phenylpropanoid moiety of the cyanidin structure is indicated by bolder print. Participating enzymes (indicated by encircled numbers): (1) DAHP-synthase; (2) dehydroquinate synthase; (3) dehydroquinate dehydratase; (4) SHK-dehydrogenase; (5) SHK kinase; (6) EPSP synthase; (7) CHA-synthase; (8) CHA-mutase; (9) PRE-aminotransferase; (10) ARO dehydratase; (11) PHE ammonia-lyase, (PAL); (12) CIN-4-hydroxylase; (13) HCCoA-ligase; (14) chalcone synthase; (I) PYR kinase; (II) PYR dehydrogenase; (III) AcCoA-carboxylase. Abbreviations: **AcCoA**, acetyl coenzyme A; **AIP**, 2-aminoindan 2-phosphonic acid.

the aromatic amino acid phenylalanine, which is one of the end products of the shikimate pathway; it is channeled into the phenylpropanoid pathway via its deamination to *trans*-cinnamic acid, catalyzed by phenylalanine ammonia-lyase (PAL). Compounds interfering *in vivo* with any enzyme of the shikimate or general phenylpropanoid pathway are, therefore, expected to reduce anthocyanin accumulation (Amrhein, 1986). The buckwheat assay described here for inhibitors of anthocyanin biosynthesis is based on the original procedures of Scherf and Zenk (1967) for the investigation of light-induced flavonoid synthesis in buckwheat seedlings.

PROCEDURE

Cultivation of Buckwheat

Buckwheat (*Fagopyrum esculentum* Moench) seeds can be obtained from many commercial seed suppliers (e.g., Samen Mauser, Zürich) and can be stored at room temperature under dry conditions for 1 to 2 years without loss of viability. They are germinated in moist vermiculite or perlite (runoff, tap water) in plastic trays (20 × 20 × 5 cm) and grown for 6 days in complete darkness at 21–22°C. Since the humidity is usually high in growth cabinets filled with trays (80% relative humidity), watering is usually required only once during the growth period.

Irradiation of Seedlings

Roots of seedlings of uniform hypocotyl length (ca. 6 cm) are removed using a razor blade or, more conveniently, a fingernail, and the seed coats are gently squeezed off the cotyledons. Derooted seedlings are collected on moist tissue paper. These operations can be carried out under normal laboratory light conditions. Batches of 20 derooted seedlings are placed with the cut hypocotyl ends in a glass or plastic scintillation vial (20 mL) containing 5 mL of 10 m*M* potassium buffer, pH 5.5, and the test compound at the desired concentration. (Use appropriate stock solutions and fill up with water to the final volume. For compounds which are barely water soluble, it is recommended to use dimethyl sulfoxide (DMSO) as solvent since the seedlings will tolerate DMSO concentrations of up to 5% without reduction of anthocyanin synthesis.) The vials are then irradiated for 18 to 24 hr with light from two 500-W quartz halogen lamps (providing a light intensity of approximately 600 μEinstein (μE) × m^{-2} × sec^{-1} at the level of the cotyledons of the

Figure 1. **(Continued) AOPP**, α-aminooxy-β-phenylpropionic acid; **ARO**, arogenate; **CHA**, chorismate; **CIN**, cinnamate; **CYA**, cyanidin; **DAHP**, 3-deoxy-D-*arabino*-sedoheptulosonate 7-phosphate; **E4P**, erythrose 4-phosphate; **EPSP**, 5-enolpyruvylshikimate 3-phosphate; **HCCoA**, hydroxycinnamoyl (= *p*-coumaroyl) coenzyme A; **MaCoA**, malonyl coenzyme A; **NAR**, naringenin chalcone; **PEP**, phosphoenolpyruvate; **PHE**, phenylalanine; **PRE**, prephenate; **PYR**, pyruvate; **SHK**, shikimate; **SHKP**, shikimate 3-phosphate; **TRP**, tryptophan; **TYR**, tyrosine.

seedlings) filtered through a 4-cm layer of running tap water for heat absorption.

Extraction of Anthocyanins

After the removal of the cotyledons, the hypocotyls are rinsed briefly with water, blotted dry, and transferred to 10 mL 1 *N* HCl in a 100-mL plastic screw cap vial to which ca. 15 g glass beads (4 mm diameter) are added. Tissue homogenization and extraction of anthocyanin are achieved by reciprocal shaking of the vials at 300 cpm. Alternatively, the hypocotyls can be homogenized with mortar, pestle, and some quartz sand, but this procedure is inconvenient for a large number of samples. The homogenates are centrifuged for 15 min at 10,000 × *g*, and the optical density of the clear supernatants at 515 nm is determined in 1-cm plastic or glass cuvettes against a control (homogenate of non-irradiated seedlings) to correct for light scattering. An optical density of 1.0 corresponds to 0.37 μmol cyanidin (glycosides) per 20 hypocotyls. Differences in the weights of individual batches can be corrected by calculating the anthocyanin content per gram fresh weight. Assays for each test compound concentration are run in duplicate. An average value for 20 hypocotyls is 2.8 g fresh weight, which gives an E_{515} of 0.9.

Equipment

Necessary equipment includes a temperature-controlled incubator for growing the seedlings in the dark, a light source and water filter (no further temperature control is required), a high-speed reciprocal shaker with 5 cm of stroke, a high-speed centrifuge, and a standard spectrophotometer (visible light only).

Herbicides and Other Inhibitors

These include glyphosate (99.9% pure; Monsanto, St. Louis, MO), AOPP (α-aminooxy-β-phenylpropionic acid; L. Andert, Ruhr University, Bochum, Germany), AIP (2-aminoindan 2-phosphonic acid; J. Zoń, Technical University Wroclaw, Wroclaw, Poland), and (*R*) 1-aminoethylphosphinate (Ciba-Geigy AG, Basel, Switzerland).

RESULTS

Figure 2 shows the inhibition of light-induced anthocyanin formation in derooted buckwheat seedlings as a function of the concentration of glyphosate, and Table 1 lists the I_{50} values determined from such curves, as well as the target enzymes for various inhibitors (see also Figure 1).

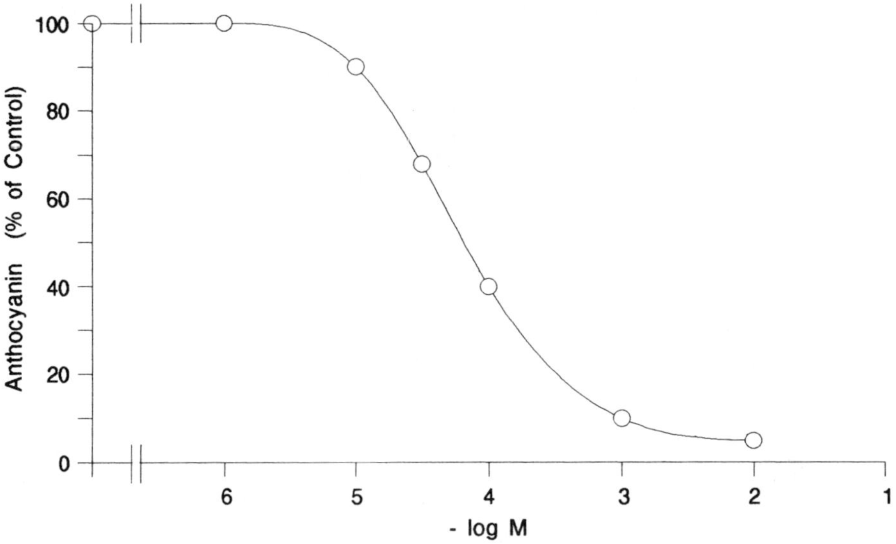

Figure 2. Inhibition of anthocyanin formation in irradiated derooted buckwheat seedlings as a function of the concentration of glyphosate. Anthocyanin formation in the untreated control was 0.11 μmol/g fresh weight.

COMMENTS

General

More than 20 enzymes are required in the synthesis of cyanidin and its glycosides from the metabolites of primary metabolism, phosphoenolpyruvate (PEP) and erythrose 4-phosphate (E4P; see Figure 1). Quite a number of the reactions in the long pathway are driven at the expense of ATP or other energy-rich compounds. Furthermore, *de novo* protein synthesis is required for the induction by light of anthocyanin synthesis. Thus, inhibitors of oxidative phosphorylation (e.g., 2,4-dinitrophenol) or of protein synthesis (e.g., cyloheximide) will inhibit anthocyanin synthesis in the buckwheat system.

Table 1. Targets of Inhibitors and I_{50} Values for Inhibition of Anthocyanin Formation in Buckwheat Hypocotyls.[a]

Compound	Target	I_{50} (μM)
(*R*)-Aminoethylphosphinate[b]	Pyruvate dehydrogenase	10
AIP	PAL	1.5
AOPP	PAL	10
Glyphosate	EPSP synthase[c]	60

[a] For abbreviations see legend to Figure 1.

[b] Aminoethylphosphinate is transaminated *in vivo* to yield acetylphosphinate, which is the actual inhibitor of pyruvate dehydrogenase (Laber and Amrhein, 1987).

[c] See Chapter 16.

However, reversal experiments with biosynthetic intermediates will indicate whether or not a general block in metabolism or a specific block in the biosynthetic pathway is produced by the inhibitor under investigation (Amrhein, 1979). Furthermore, the accumulation of specific intermediates of the pathway in inhibitor-treated tissue will provide a clue for the site of the block: e.g., shikimate and shikimate 3-phosphate accumulate in the presence of glyphosate; the content of free phenylalanine increases drastically in tissues treated with AOPP or AIP; and pyruvate, alanine, and other pyruvate-derived metabolites accumulate in the presence of aminoethylphosphinate (see Figure 1). The transamination product of aminoethylphosphinate, acetylphosphinate (a pyruvate analogue), was in fact rationally designed as a potential herbicide, but its activity *in vivo* is very low (Baillie et al., 1988).

Practical

While it is somewhat tedious to remove the seed coats from the cotyledons of the seedlings for the assay, their removal is encouraged because the cotyledons will expand in the light and, thus, increase the uptake of the test compounds and their transport up the hypocotyl with the transpiration stream. I_{50} values for inhibition of anthocyanin synthesis are generally higher when excised hypocotyls are floated on inhibitor solutions. Uptake via the hypocotyl epidermis is apparently less efficient in delivering the inhibitors to their sites of action than uptake via the cut hypocotyl ends.

The buckwheat system also can be conveniently used to test for the phloem mobility of inhibitors. For this test, *intact* etiolated seedlings (i.e., with the roots attached but with seed coats removed) are used. A 10-μL drop of inhibitor solution containing 0.1% Triton X-100 is carefully placed between the folded cotyledons with the aid of a micropipette. A phloem-mobile compound such as glyphosate will be transported down the hypocotyl and will cause inhibition of light-induced anthocyanin accumulation, while an immobile compound will not affect pigment synthesis.

REFERENCES

Amrhein, N., Biosynthesis of cyanidin in buckwheat hypocotyls. *Phytochemistry,* 18, 585, 1979.

Amrhein, N., Specific inhibitors as probes into the biosynthesis and metabolism of aromatic amino acids, in *The Shikimic Acid Pathway,* Conn, E. E., Ed., Plenum Press, New York, 1986, p. 83.

Baillie, A. C., Wright, K., Wright, B. J. and Earnshaw, C. G., Inhibitors of pyruvate dehydrogenase as herbicides. *Pestic. Biochem. Physiol.,* 30, 103, 1988.

Laber, B. and Amrhein, N., Metabolism of 1-aminoethylphosphinate generates acetylphosphinate, a potent inhibitor of pyruvate dehydrogenase. *Biochem. J.,* 248, 351, 1987.

Scherf, H. and Zenk, M. H., Der Einfluss des Lichtes auf die Flavonoidbiosynthese und die Enzyminduktion bei *Fagopyrum esculentum* Moench. *Z. Pflanzenphysiol.,* 57, 401, 1967.

CHAPTER 16

EPSP Synthase: Isolation and Assay

Nikolaus Amrhein

INTRODUCTION

The enzyme 5-enolpyruvylshikimate 3-phosphate (EPSP) synthase (3-phosphoshikimate-1-carboxyvinyl transferase, EC 2.5.1.19) catalyzes the sixth and penultimate step in the prechorismate pathway. Chorismic acid is the precursor of the aromatic amino acids phenylalanine, tyrosine, and tryptophan, as well as of other important aromatic metabolites (see Figure 1 of Chapter 15). EPSP synthase is the biological target of glyphosate (*N*-phosphonomethylglycine; Figure 1), the active component of ROUNDUP® herbicide, and the biochemistry and molecular biology of this enzyme from bacterial, fungal, and higher plant sources have been investigated in considerable detail (Sikorski et al., 1991). EPSP synthase catalyzes the unusual and reversible transfer reaction of the enolpyruvyl (carboxyvinyl) group from phosphoenolpyruvate (PEP) to the 5-hydroxyl group of shikimate 3-phosphate (S3P), producing EPSP and inorganic phosphate (Pi) (Figure 1). EPSP synthase can be assayed in either the forward or the reverse direction. In both cases, a substrate (S3P or EPSP), unavailable commercially, has to be prepared. Since EPSP must be prepared enzymatically from S3P, its synthesis requires steps in addition to the synthesis of S3P and is therefore not described here. For the forward assay, at least three options exist:

0-87371-539-X/93/$0.00 + $.50

1. Residual PEP is measured by coupling of the pyruvate kinase and lactate dehydrogenase reactions.
2. Formation of EPSP is coupled to the chorismate synthase (and, for increased sensitivity, to the anthranilate synthase) reaction.
3. Pi released during the reaction is measured by the sensitive malachite green assay.

The third option is described here. Detailed protocols for the extraction, purification, and assay of EPSP synthases from *Escherichia coli* and *Pisum sativum,* as well as a full set of references to the various assay procedures, are given by Lewendon and Coggins (1987) and Mousdale and Coggins (1987), respectively. The procedures described here are mainly based on the work of Steinrücken et al. (1986).

PROCEDURE

Preparation of S3P

S3P is isolated from the growth medium of the *Klebsiella pneumoniae* ATCC 25597 strain, which is deficient in EPSP synthase. A detailed, updated protocol was recently provided by Coggins et al. (1987).

EPSP Synthase Assay

The assay mixture contains 50 m*M* HEPES 4-(2-hydroxylethyl)-1-piperazineethanesulfonic acid) (HEPES)/KOH, 1 m*M* S3P, 1 m*M* PEP, and 0.1 m*M* ammonium heptamolybdate, pH 7.0, in a final volume of 100 μL. After preincubation (5 min, 30°C), the reaction is started by the addition of enzyme. After an appropriate incubation period at 30°C, the reaction is terminated by the addition of 1 mL phosphate reagent (0.33 g malachite green, 1.03 g ammonium heptamolybdate, 20 mL Tergitol NPX in 1 L 1 *N* HCl) followed by 0.2 mL of 34% (w/v) trisodium citrate after 1 min. After 90 min, the optical density at 660 nm of the samples is determined in halfmicro plastic cuvettes. For assay blanks, 10 m*M* glyphosate is added to the reactions.

EPSP Synthase Extraction and Purification

Plant material or cultured plant cells (100 to 500 g fresh weight) are frozen in liquid N_2 and ground to a powder. The frozen material is allowed to thaw in four volumes of buffer A (20 m*M* HEPES/KOH, 10 m*M* glutathione, 0.1 m*M* EDTA, pH 7.0) containing 1% (w/v) polyvinylpolypyrrolidone (PVP) and stirred at the temperature of melting ice for 30 min. The crude homogenate is filtered through Miracloth and clarified by centrifugation at 20,000 × *g* for 20 min. The resulting supernatant is fractionated with ammonium sulfate, and the 40 to 65% saturated fraction is collected and centrifuged at 20,000 × *g* for 10 min. The precipitate is redissolved in a small volume of buffer

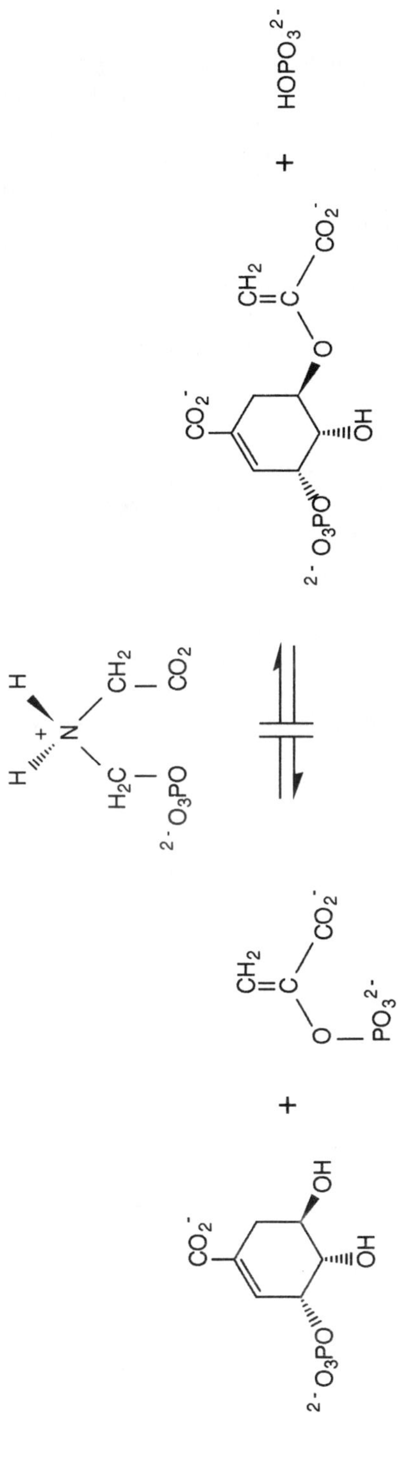

S3P PEP Glyphosate EPSP Pi

Figure 1. Reaction catalyzed by EPSP synthase, and structure of glyphosate.

B (10 m*M* 2-(*N*-morpholino)ethanesulfonic acid (MES), 10 m*M* glutathione, 10% (v/v) glycerol, 10 μ*M* sodium tungstate, adjusted to pH 6.5 with KOH). After desalting on a Sephadex G-25 column (5 × 60 cm), the protein fraction is applied to a Whatman phosphocellulose column (5 × 15 cm) equilibrated with buffer B. The column is washed extensively with the same buffer until no protein is detected in the eluate (OD_{260}), and EPSP synthase is then selectively eluted (flow rate = 1 mL · min^{-1}) with 50 mL of buffer B, but with an addition of S3P and glyphosate, each at a concentration of 0.1 m*M*. Fractions containing EPSP synthase activity are pooled and stored at −18°C. Prior to use, the preparation is dialyzed overnight against a 1000-fold excess of 20 m*M* HEPES, 10 m*M* glutathione, and 10% (v/v) glycerol, the pH adjusted to 7.1 with KOH.

For tissue extracts with high phosphatase and/or low EPSP synthase activities, chromatography on DEAE cellulose is recommended before cellulose chromatography. For this, the 40 to 65% ammonium sulfate precipitate (see above) is redissolved in buffer A containing 10% glycerol and applied to the Sephadex G-25 column (see above; equilibrated with buffer A) which is connected directly to a DEAE cellulose (Whatman DE-52) column (1.5 × 20 cm) equilibrated with buffer A. After passing of the protein peak into the DEAE column, the latter is disconnected and eluted with a 0 to 0.2 *M* KCl linear gradient prepared in buffer A. Fractions containing EPSP synthase activity are pooled; proteins are precipitated by addition of ammonium sulfate to 80% saturation and redissolved in buffer B. For phosphocellulose chromatography, continue as described above.

Equipment

Necessary equipment includes a high-speed refrigerated centrifuge, chromatography columns, fraction collector, cold room or cabinet, and a spectrophotometer (UV/visible).

Herbicides

The herbicide used was glyphosate, free acid (99.9% pure; Monsanto, St. Louis, Missouri).

RESULTS

Cultured plant cells adapted to growth in the presence of high glyphosate concentrations (1 to 10 m*M*) are a good source of EPSP synthase, since they overproduce the enzyme so that it may constitute 1 to 2% of the soluble protein of the cells (see Steinrücken et al., 1986). Table 1 shows that the enzyme can be purified from such cells to high specific activity and with good yield, essentially by ammonium sulfate fractionation and phosphocellulose affinity chromatography. The procedure requires 2 days and yields a stable preparation. Purification of EPSP synthase from nonadapted cells requires (at least) one additional step (DEAE-cellulose), and specific activity and yield of the preparation are lower (Table 1). Nevertheless, satisfactory purification, as judged by sodium dodecyl sulfate polyacrylamide gel electrophoresis (SDS-PAGE), can be achieved by this method for EPSP synthase from both sources (Figure 2). Using routine protocols for the analysis of enzyme inhibition kinetics, glyphosate is found to inhibit EPSP synthase from both *Corydalis sempervirens* cell cultures competitively with respect to PEP (K_i = 0.32 μ*M*).

COMMENTS

EPSP Synthase Assay

The stopped assay using malachite green for the quantitation of inorganic phosphate released during the EPSP synthase reactions is quite sensitive, but one may encounter difficulties with residual phosphatase activity in crude extracts which is not fully inhibited by molybdate. For assay blanks, 10 m*M* glyphosate is therefore added to the reactions to determine the release of phosphate which is not due to EPSP synthase activity.

EPSP Synthase Isolation

Compared with microorganisms, plant tissues are poor sources of EPSP synthase activity (Mousdale and Coggins, 1987). EPSP synthase-overproducing plant cell cultures adapted to growth in the presence of glyphosate are therefore suitable sources of the enzyme (Table 1). The phosphocellulose affinity chromatography which exploits the highly specific formation of the EPSP synthase·S3P·glyphosate dead-end complex (Sikorski et al., 1991) for elution of the bound enzyme is extremely efficient provided that phosphatase activity can be sufficiently inhibited by the added tungstate and/or be removed by prior DEAE-cellulose chromatography (Table 1). For kinetic studies and

Table 1. Purification of EPSP Synthase from (A) Glyphosate-Adapted and (B) Nonadapted Cultured Cells of *Corydalis sempervirens*[a]

	Protein (mg)		Specific Activity (nkat · mg^{-1} protein)		Yield (%)	
	A	B	A	B	A	B
Crude extract	1287	1420	10.4	0.36	100	100
Ammonium sulfate precipitation	601	725	19.4	0.66	88	92
Sephadex G-25	543	682	16.7	0.59	68	84
DEAE-Cellulose	—	45	—	5.0	—	56
Sephadex G-25	—	50	—	4.65	—	45
Phosphocellulose	13.5	0.38	584	516	59	38

[a] Data of D. Johänning and N. Amrhein.

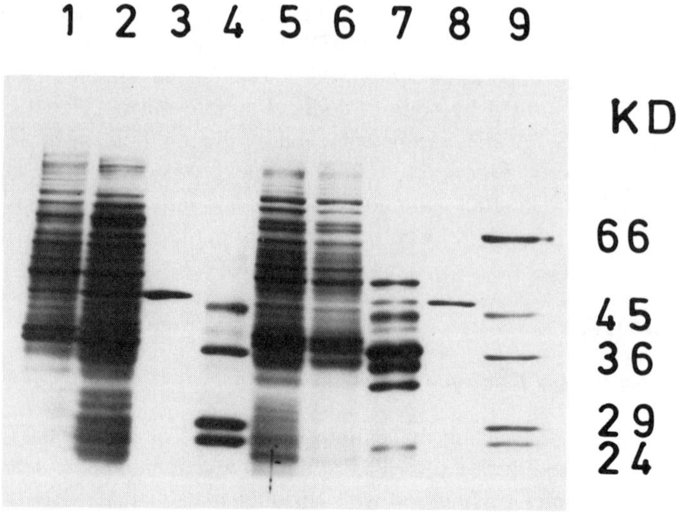

Figure 2. SDS-PAGE of EPSP synthase from glyphosate-adapted (lanes 1–3) and nonadapted (lanes 5–8) cultured cells of *Corydalis sempervirens*. Lane 1: crude extract, 50 μg protein. Lane 2: Sephadex G-25, 50 μg protein. Lane 3: phosphocellulose, 7.5 μg protein. Lane 5: crude extract, 50 μg protein. Lane 6: Sephadex G-25, 50 μg protein. Lane 7: DEAE-cellulose, 25 μg protein. Lane 8: phosphocellulose, 5 μg protein. Lanes 4 + 9: molecular weight markers. Data of D. Johänning and N. Amrhein.

the evaluation of various potential inhibitors, it is essential to have an enzyme preparation, at least partially purified, which can be stored for longer periods; the protocol described here is suitable for this purpose. Mousdale and Coggins (1987) have described a small-scale, rapid purification of EPSP synthase from chloroplast preparations which may be useful in screening different plant species for the sensitivity of EPSP synthase to potential inhibitors such as glyphosate.

EPSP Synthase Inhibitors

Besides glyphosate, only its *N*-hydroxy- and *N*-amino-analogues inhibit EPSP synthase with potency similar to that of the lead compound, while *N*-alkyl derivatives are totally inactive (Sikorski et al., 1991). For all practical purposes, glyphosate is still the only member of its class.

REFERENCES

Coggins, J. R., Boocock, M. R., Chaudhuri, S., Lambert, J. M., Lumsden, J., Nimmo, G. A., and Smith, D. D. S., The *arom* multifunctional enzyme from *Neurospora crassa*. *Methods Enzymol.*, 142, 325, 1987.

Lewendon, A. and Coggins, J. R., 3-Phosphoshikimate 1-carboxyvinyltransferase from *Escherichia coli*. *Methods Enzymol.*, 142, 342, 1987.

Mousdale, D. M. and Coggins, J. R., 3-Phosphoshikimate 1-carboxyvinyltransferase from *Pisum sativum. Methods Enzymol.*, 142, 348, 1987.

Sikorski, J. A., Anderson, K. S., Cleary, D. G., Miller, M. J., Pansegrau, P. D., Ream, J. E., Sammons, R. D., and Johnson, K. A., An enzyme-targeted herbicide design program based on EPSP synthase: chemical mechanism and glyphosate inhibition studies, in *Chemical Aspects of Enzyme Biotechnology: Fundamentals,* Baldwin, T. D., Rausel, F. M. and Scott, A. I., Eds., Plenum Press, New York, 1991, p. 23.

Steinrücken, H. C., Schulz, A., Amrhein, N., Porter, C. A., and Fraley, R. T., Overproduction of 5-enolpyruvylshikimate-3-phosphate synthase in a glyphosate-tolerant *Petunia hybrida* cell line. *Arch. Biochem. Biophys.*, 244, 169, 1986.

CHAPTER 17

Investigations on Herbicidal Acetolactate Synthase Inhibitors Using Valine-Sensitive and Valine-Insensitive *Escherichia coli* Strains

Peter Babczinski

INTRODUCTION

Inhibition of essential amino acid biosynthesis in plants is one of the most successful biochemical principles of herbicidal mode of action. Impairment of the biosynthesis of branched-chain L-amino acids valine, leucine, and isoleucine by inhibitors of acetolactate synthase (ALS [acetohydroxyacid synthase, AHAS]; E.C. 4.1.3.18) led to the discovery of the most active group of herbicides to date. Some of these compounds are active in the field in grams per hectare amounts ("tablespoon per acre") and exhibit exceptionally low inhibition constants on isolated enzyme preparations (nanomolar range).

At least 13 structurally diverse classes of ALS-herbicides have been described in the past (see, e.g., Babczinski and Zelinski, 1991, for structural formulas), and rapid and convenient assay systems are helpful in chemical optimization work. This chapter describes a microbial test system using an *Escherichia coli* mutant for the discovery of herbicides. It facilitates the search for new lead compounds and allows the quantitative evaluation of the herbicidal potency of chemicals by determining I_{50} or MIC values. In addition, reversal experiments can be performed in order to exclude artificial inhibitory

0-87371-539-X/93/$0.00 + $.50

effects of test chemicals. Moreover, an elegant thin-layer biogram technique provides the opportunity to investigate mixtures of chemicals.

PROCEDURE

Chemicals

Analytical-grade chemicals and biochemicals from usual sources were used. Porcine liver carboxylesterase was from Sigma (St. Louis, Missouri, U.S.) and BiTek agar and antibiotic medium 3 were from Difco (Detroit, Michigan, U.S.).

The different ALS inhibitors used were bensulfuron (methyl-2-((((((4, 6-dimethoxy-2-pyrimidinyl)-amino)-carbonyl)-amino)-sulfonyl)-methyl)-benzoate), chlorsulfuron (2-chloro-*N*-(((4-methoxy-6-methyl-1,3,5-triazine-2-yl)-amino)-carbonyl)-benzenesulfonamide), imazapyr (2-(4,5-dihydro-4-methyl-4-(1-methylethyl)-5-oxo-1H-imidazol-2-yl)-3-pyridinecarboxylic acid), metsulfuron (methyl-2-(((((4-methoxy-6-methyl-1,3,5-triazine-2-yl)-amino)-carbonyl)-amino)-sulfonyl)-benzoic acid), nicosulfuron (2-(((((4,6-dimethoxy-2-pyrimidinyl)-amino)-carbonyl)-amino)-sulfonyl)-*N,N*-dimethyl-3-pyridinecarboxamide), primisulfuron (methyl-2-(((((4,6-bis(difluoromethoxy)-2-pyrimidinyl)-amino)-carbonyl)-amino)-sulfonyl)-benzoic acid), pyrazosulfuron (ethyl-5-(((((4,6-dimethoxy-2-pyrimidinyl)-amino)-carbonyl)-amino)-sulfonyl)-1-methyl-1H-pyrazole-4-carboxylic acid), sulfometuron (methyl-2-(((4,6-dimethyl-2-pyrimidinyl)-aminocarbonyl)-aminosulfonyl)-benzoic acid), thiameturon (methyl-3-(((((4-methoxy-6-methyl-1,3,5-triazine-2-yl)-amino)-carbonyl)-amino)-sulfonyl)-2-thiophenecarboxylic acid), and 2-(4′,6′-dimethoxypyrimidine-2′-yl)oxybenzoic acid methyl ester. All amino acids were the L-isomers unless stated otherwise.

Cultivation

E. coli wild-type strains 14 and K12 were grown at 37°C in antibiotic medium 3 or in medium containing potassium dihydrogen phosphate, 3 g/L; dipotassium hydrogen phosphate, 7 g/L; sodium chloride, 0.1 g/L; ammonium sulfate, 1 g/L; magnesium sulfate heptahydrate, 0.1 g/L; glucose, 500 g/L; and agar, 20 g/L. Precultures were kept in antibiotic medium 3 containing glycerol (600 g/L). *E. coli* strain FD 1062 (kindly provided by Prof. R. P. Lawther, University of South Carolina, Columbia) was grown in antibiotic medium 3 containing kanamycin (25 mg/L) at 30°C or in medium containing salts, glucose, and agar as described for *E. coli* wild-type strains but at 24°C. Additionally, kanamycin (25 mg/L) and 18 proteinergic amino acids (each 150 mg/L) with the omission of leucine and isoleucine were present. Precultures were kept as in the case of the wild types; in addition, kanamycin (25 mg/L) was present.

For agar diffusion tests, holes were stamped in solidified *E. coli* agar (in quadratic Nunc plates), and 100 μL of inhibitor solution in methanol were applied.

For the thin-layer bioassay diagrams, precoated preparative silica plates placed in Nunc plates were overlayed with *E. coli* agar and incubated for 2 days at 24°C. Inhibition zones were located by direct visual judgment under ultraviolet (UV) light or by stripping off the agar layer from the silica plate and subsequent visual analysis.

Equipment

For this assay as basic equipment a rotary shaker (30°C and 37°C) and an oven (24°C and 37°C) are required. Precultures should be kept at −20°C.

Sources

E. coli strain 14 was obtained from Bayer AG, Health Care, Dr. K.-G. Metzger, Geb. 405, Aprather Weg, 5600 Wuppertal 1, Germany.

E. coli strain FD 1062 was obtained from the University of South Carolina, Department of Biological Sciences, Prof. Dr. R. P. Lawther, Columbia, S C 29208.

RESULTS

Herbicides in *E. coli* Wild-Type Strain Assays

Growth is not inhibited (Table 1) by the sulfonylurea herbicide chlorsulfuron in up to millimolar concentrations in strain K12 (the *ilv*G gene product ALS isoenzyme II is not expressed; only ALS II is the target for sulfonylureas in bacteria [LaRossa and Schloss, 1984] or in strain 14 (which is not a K strain; valine-resistant ALS II as well as valine-sensitive ALS I and III are expressed) if grown in minimal medium (organic C, inorganic N). In the presence of valine, however, ALS-inhibitory herbicides reduce growth resulting from a starvation for isoleucine. Valine cannot be replaced by any of 19 other proteinergic amino acids. Growth inhibition by the mixture of chlorsulfuron plus valine can be specifically suppressed by L-isoleucine.

Herbicides in *E. coli* FD 1062 Assays

Using *E. coli* mutant strain FD 1062 (only the *ilv*G gene product valine-resistant ALS II is expressed, but none of the valine-sensitive isoenzymes), growth inhibition is already observed in the absence of valine (Table 1), and this again can be specifically overcome by the addition of either isoleucine or its specific biosynthetic precursor, ketomethylvaleric acid. Neither valine- nor leucine-specific ketoacids, nor the substrates of ALS isoenzyme I, II, and

Table 1. Phenotypes of *E. coli* Wild-Type Strains 14 and K12 and Strain FD 1062.

Condition	Growth Response[a] Strain 14	Strain K12
Valine	+	−
19 other amino acids	+	+
Chlorsulfuron (I)	+	+
I plus valine	−	−
I plus 19 other amino acids	+	+
I plus valine plus isoleucine	+	+
I plus valine plus 18 other amino acids	−	−
I plus valine plus D-isoleucine	−	−

Condition	Strain FD 1062
Valine	+
19 other amino acids	+
Chlorsulfuron (I)	−
I plus isoleucine	+
I plus 19 other amino acids	−
I plus D-isoleucine	−
I plus valine plus isoleucine	−
I plus ketomethylvaleric acid	+
I plus ketoisocaproic acid or ketoisovaleric acid	−
I plus pyruvate or threonine or α-keto-butyric acid	−

[a] +, indicates growth; −, indicates growth inhibition.
Source: Adapted from Babczinski and Zelinski (1991). With permission.

III, pyruvate, threonine, or α-ketobutyric acid, can reverse growth inhibition by chlorsulfuron.

Using this agar plate diffusion assay, MIC values of large numbers of chemicals can be determined in dilution series. Some examples involving sulfonylurea herbicides are given in Table 2. The MIC values vary over a range from 5 n*M* to 0.2 m*M* (chlorsulfuron: 11.2 μ*M*). The inhibitory activities of all of these compounds can be reverted by the inclusion of isoleucine in the agar diffusion assay and, thus, a specific interference with the biosynthesis of this branched-chain amino acid can be demonstrated.

The bacterial test system can be adapted to a thin-layer chromatogram bioassay for investigations on mixtures of chemicals (especially those with contaminants of unknown chemical structures). Developed preparative thin-layer plates (silica gel, F_{254}) were coated with *E. coli* FD 1062 agar (1 to 2 mm thick) and incubated for 2 days at 24°C. Under UV light the spots of growth inhibition look fluorescent green within a completely quenched background (bacteria, but not agar, quench fluorescence). Due to diffusion of the substances from the plate into the agar, the original UV quench spots disappear and do not interfere with growth inhibition zones. The bioassay performed with the herbicide sulfometuron is more sensitive than UV detection by approximately a factor of 50 (Figure 1; in this case the test compound is shown clearly to have two contaminants which most probably reflect ALS inhibitors; the main product obviously is inactive).

COMMENTS

This simple and quick microbial test system for herbicidal compounds offers the possibility to screen ALS inhibitors, to reveal the biochemical target, and to determine MIC values (or I_{50} values) for their specific activity. These *in vitro* values do not necessarily reflect the amounts of ingredients needed in greenhouse experiments or K_i values obtained with isolated plant enzyme, although correlations between MIC values and greenhouse data for separate subclasses within a particular synthetic ALS program are quite acceptable (correlation factor $r > 0.9$).

Certainly, compounds with different biochemical modes of action (e.g., antibiotics, respiration inhibitors, etc.) induce inhibition zones in this agar diffusion assay which might imitate ALS activity. Reversion experiments with isoleucine therefore are of importance to exclude such artifacts.

Unexpectedly, herbicidal ALS-inhibitory imidazolinones (imazapyr; Shaner et al., 1984) as a whole group of compounds are inactive against *E. coli* FD 1062 or the wild-type strains in the presence of valine, although they exert activity on isolated bacterial acetolactate synthase (Schloss et al., 1988; Babczinski and Zelinski, 1991). Obviously, restricted uptake and/or metabolism is the reason for this remarkable phenomenon.

Herbicidal ALS-inhibitory pyrimidylsalicylate esters (Hawkes, 1989) are only active after ester cleavage with carboxylesterase. This can be demonstrated easily using the agar diffusion assay, where only in the presence of the enzyme in the substance hole are inhibition zones obtained. In this respect, this result is inverse to the sulfonylurea herbicide sulfometuron, where the ester form is active whereas the free acid is weaker by three orders of magnitude.

Table 2. MIC Values of Selected Sulfonylureas against *E. coli* FD 1062.[a]

Compound	MIC (μ*M*)
Nicosulfuron	0.005
Metsulfuron	0.3
Thiameturon	2.6
Bensulfuron	4.9
Sulfometuron	5.5
Chlorsulfuron	11.2
Pyrazosulfuron	19.3
Primisulfuron	267

[a] MIC, minimal inhibitory concentration.

Source: Adapted from Babczinski and Zelinski (1991). With permission. © Society of Chemical Industry.

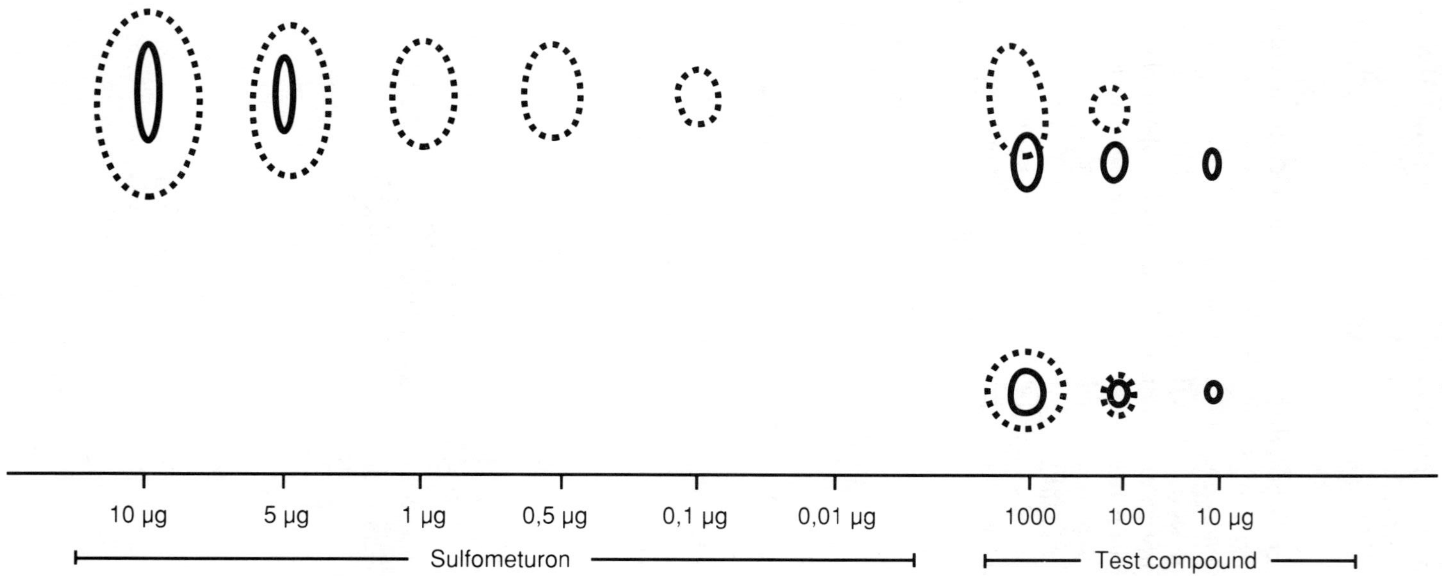

Figure 1. Thin-layer chromatography biogram with *Escherichia coli* FD 1062. Solid circles represent quenched fluroescence (marked before agar application); dashed circles indicate inhibition zones of bacterial growth (marked after bacterial growth). (After Babczinski, P. and Zelinski, T., *Pestic. Sci.*, 31, 305, 1991. With permission. © Society of Chemical Industry.)

REFERENCES

Babczinski, P. and Zelinski, T., Mode of action of herbicidal ALS-inhibitors on acetolactate synthase from green plant cell cultures, yeast, and *Escherichia coli*. *Pestic. Sci.*, 31, 305, 1991.

Hawkes, T. R., Studies of herbicides which inhibit branched chain amino acid biosynthesis, in *Prospects for Amino Acid Biosynthesis Inhibitors in Crop Protection and Pharmaceutical Chemistry*, Copping, L. G., Dalziel, J., and Dodge, A. D., Eds., British Crop Protection Council, Farnham, Surrey, 1989, p. 131.

LaRossa, R. A. and Schloss, J. V., The sulfonylurea herbicide sulfometuron methyl is an extremely potent and selective inhibitor of acetolactate synthase in *Salmonella typhimurium*. *J. Biol. Chem.*, 259, 8753, 1984.

Schloss, J. V., Ciskanik, L. M., and Van Dyk, D. E., Origin of the herbicide binding site of acetolactate synthase. *Nature*, 331, 360, 1988.

Shaner, D. L., Anderson, P. C., and Stidham, M. A., Imidazolinones (potent inhibitors of acetohydroxyacid synthase). *Plant Physiol.*, 76, 545, 1984.

CHAPTER 18

Assay of Acetolactate Synthase from Higher Plants

Jörg Durner and Valérie Gailus

INTRODUCTION

Acetolactate synthase (ALS, EC 4.1.3.18; also referred to as acetohydroxy acid synthase, AHAS) catalyzes the first common step in the biosynthetic pathway of the essential branched-chain amino acids valine, leucine, and isoleucine. The enzyme can catalyze either the formation of acetohydroxybutyrate from pyruvate and γ-oxobutyrate or the synthesis of acetolactate from two molecules of pyruvate. The enzyme has raised strong interest since several structurally unrelated classes of herbicides — i.e., the sulfonylureas and the imidazolinones — have been shown to specifically inhibit ALS. These compounds exhibit strong herbicidal potency combined with low mammalian toxicity, probably due to the fact that mammals lack ALS.

Due to extreme lability purification of the eukaryotic ALS enzymes is very difficult (Durner and Böger, 1988). In this chapter, a procedure to obtain ALS with a high specific activity from higher plants is presented and some details with respect to inhibition are given.

0-87371-539-X/93/$0.00 + $.50

PROCEDURE

Cultivation

Seeds of barley (*Hordeum vulgare* L.) were placed onto a nylon net, covered with filter paper, wetted with distilled water, and allowed to germinate in a dark chamber at 23 to 26°C. After germination (24 to 36 hr), roots were immersed in a medium containing 2 g/L KNO_3, 34 mg/L $CaSO_4$, and 0.25 mg/L Na_2MoO_4, pH 7.0. Three days after germination, etiolated shoots were harvested, washed in distilled water, collected, and stored at 70 K for later use.

Seeds of corn (*Zea mays* L.) were soaked in water and cultivated in moistened vermiculite in a dark chamber at 33°C. Four-day-old shoots were treated as described for barley.

Equipment

For this assay a heating block, a centrifuge (e.g., Sorvall RC-2), a lab centrifuge (Eppendorf), a blender, and a standard spectrophotometer are required. As an alternative to the described cultivation method, the etiolated barley can be grown in moistened vermiculite.

Herbicides

The herbicides used were chlorsulfuron, 2-chloro-*N*-[[4-methoxy-6-methyl-1,3,5-triazin-2-yl)-amino]carbonyl]benzenesulfonamide (Glean), and imazaquin, 2-(4,5-dihydro-4-methyl-4-(1-methylethyl)-5-oxo-1*H*-imidazol-2-yl)-3-quinolinecarboxylic acid (Scepter).

Enzyme Extraction

All operations were carried out at 4°C. Frozen shoots, usually 100 to 150 g, were homogenized in 400 to 600 mL of standard buffer (50 m*M* tris (hydroxymethyl)aminoethane (Tris)/HCl pH 7.5, 5 m*M* $MgCl_2$, 1 m*M* EDTA, and 15% (v/v) glycerol) containing 1 m*M* L-leucine, 1 m*M* L-valine, 0.1 m*M* flavin adenine dinucleotide (FAD), and 1 m*M* dithiothreitol (DTT). Each sample of the shoots was homogenized in a blender four times, with a 10- to 20-sec period. Cell debris was removed by centrifugation for 7 min at 20,000 × *g*. Solid ammonium sulfate was added to the supernatant (45% saturation, 2.76 g/10 mL) and the precipitate collected by centrifugation for 15 min at 27,000 × *g*. The precipitate was resuspended in standard buffer and desalted on Sephadex G-25 PD-10 minicolumns equilibrated with standard buffer. This fraction was divided into aliquots and could be stored at 70 K for later use. Specific activities were between 1 and 2 nmol acetoin per minute per milligram, depending on the quality of the plant material used.

Enzyme Assay

The standard assay mixture contained 32 m*M* Na-phosphate, pH 6.5, 40 m*M* Na-pyruvate, 0.5 m*M* thiamine pyrophosphate (TPP), 1 m*M* $MnCl_2$, and 50 μ*M* FAD, plus protein extract (1 to 1.5 mg), in a final volume of 0.5 mL. After incubation for 30 min at 30°C, the reaction was stopped with 0.25 mL 3 *M* H_2SO_4. The tubes were heated at 60°C for 12 min, and acetoin was determined by the method of Westerfeld (1945). After addition of 400 μL of 1-naphthol (100 m*M* stock in 25% NaOH) and 150 μL creatin (30 m*M* stock), color development was for 15 min at 60°C. Under these conditions 100 nmol/ 1.3 mL acetoin were equivalent to an absorbance of 1.5 at 530 nm. Protein was determined according to the method of Bradford.

RESULTS

Specifity of the ALS Assay

Using the pipet scheme listed above, a background absorbance of 0.08 to 0.1 (530 nm) could be observed. This background is not due to any enzymatic reaction, but to additives such as glycerol or FAD in the enzyme extract. A puzzling feature remains the amount of acetoin present in the assay even without decarboxylation of the produced acetolactate by acidification. However, there is strong evidence that at least part of the ALS-catalyzed acetolactate undergoes nonenzymatic decarboxylation, forming acetoin. The amount of acetoin depends strongly on feedback inhibition and inhibition by chlorsulfuron.

Crude Plant Extract *vs.* Purified Enzyme

Since highly active inhibitors like chlorsulfuron are utilized at low concentrations in enzyme assays, part of the compound may be inactivated by nonspecific binding to contaminating proteins. Therefore, two different purification grades of ALS from barley and from corn have been assayed for inhibition by chlorsulfuron and imazaquin (Table 1). In this study, purified barley enzyme does not imply a homogeneous preparation of ALS by the criteria of sodium dodecyl sulfate polyacrylamide gel electrophoresis (SDS-PAGE), as has been published previously (Durner and Böger, 1988), but indicates a more than 2000-fold enrichment with ALS as the major band on SDS-PAGE, representing more than 55% of the total protein. Protein in the assay was 1.2 mg for the crude extracts and 0.5 μg for the purified barley enzyme (partially purified corn enzyme, 0.1 mg). In all experiments, imazaquin and especially chlorsulfuron were found to be comparable potent inhibitors of ALS activity in crude extracts and (partially) purified ALS.

Table 1. I_{50} Values for Inhibition of Acetolactate Synthase from Barley and Corn by Chlorsulfuron and Imazaquin.[a]

Preparation	Chlorsulfuron (n*M*)	Imazaquin (μ*M*)
Barley		
Crude extract	34	6.0
Purified enzyme[b]	33	3.8
Corn		
Crude extract	26	9.5
Partially purified enzyme[b]	18	12.0

[a] Specific activities were 1.2 nmol acetoin per minute per milligram for the crude extract and 2.9 μmol min^{-1} mg^{-1} for the purified enzyme from barley (corn: 1.0 and 25 nmol acetoin per minute per milligram). The I_{50} value for inhibition is defined as the concentration of inhibitor which inhibits acetolactate synthase activity 50% in a 30-min standard assay. Calculation was done by regression lines of double-reciprocal plots ($r^2 > 0.99$). Values are the average of plots from at least two separate experiments.

[b] Purification was carried out as described in Durner and Böger (1988) and Durner et al. (1991).

Slow-Binding Inhibition

Figure 1 proves the time-dependent inhibition of ALS by chlorsulfuron. This means that interaction with the enzyme starts with an initial inhibition, followed by a time-dependent increase in potency leading to a final steady-state level of inhibition (Schloss, 1984). The transition between initial and final inhibition depends strongly on the applied inhibitor concentration (50 m*M*, 30 min; 5 m*M*, 90 min).

Mode of Action of ALS Inhibitors

A precise analysis of inhibition of barley ALS as a function of pyruvate concentration is shown by Figures 2A and B. Double-reciprocal plots proved to be noncompetitive and uncompetitive with respect to pyruvate for the initial inhibition by chlorsulfuron and imazaquin, respectively (initial inhibition was approximated by using a 4-min assay). K_i values were 68 n*M* for chlorsulfuron and 10 μ*M* for imazaquin. The values for the steady-state were determined as 3 n*M* for chlorsulfuron and 0.55 μ*M* for imazaquin (data not shown; Durner et al., 1991).

COMMENTS

A fast procedure is presented to obtain crude extracts from plants containing ALS activity. This preparation is suitable to determine the relative potency of ALS inhibitors. "Relative" means that due to time-dependent inhibition of fixed-time assay as described will lead to an apparent I_{50} value and not to substrate- and time-independent K_i values. However, in most cases this method

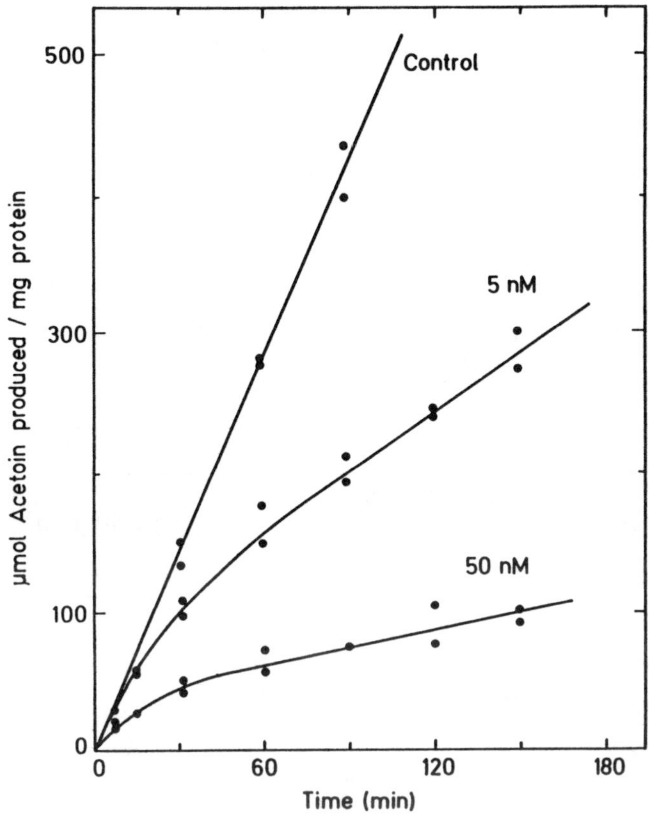

Figure 1. Time-dependent inhibition of barley ALS by chlorsulfuron. The amount of acetoin (acetolactate) produced is plotted against the incubation time. Protein in the assay was 0.1 μg with a specific activity of 4.6 μmol min^{-1} mg^{-1}. (From Durner, J., Gailus, V., and Böger, P., *Plant Physiol.*, 95, 1144–1149, 1991. With permission.)

will allow for a comparison of a new compound with standards such as chlorsulfuron or imazaquin. For a detailed analysis of the mode of action of a compound, either initial or steady-state inhibition should be examined. These experiments require at least a 100-fold enrichment of ALS in the preparation used. Partial purification can be achieved by a two-step procedure of anion-exchange and size-exclusion chromatography (Durner and Böger, 1988).

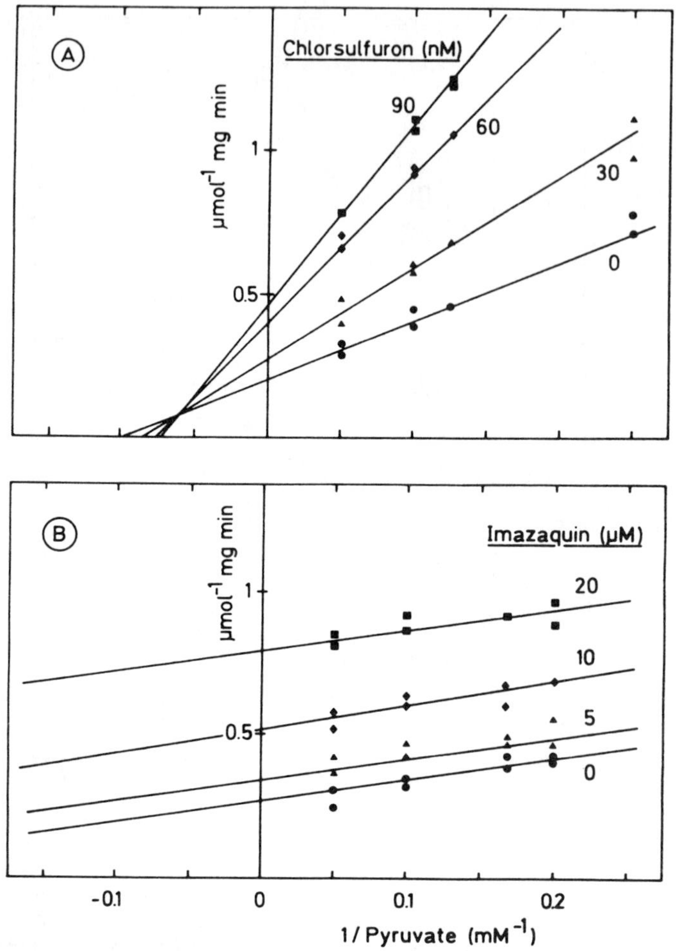

Figure 2. Initial inhibition of purified barley ALS by (A) chlorsulfuron and (B) imazaquin in the presence of different pyruvate concentrations. Data refer to a 0- to 4-min assay with specific activities of the controls of 2.8 μmol min^{-1} mg^{-1} (regression coefficient > 0.99).

REFERENCES

Durner, J. and Böger, P., Acetolactate synthase from barley (*Hordeum vulgare* L.): purification and partial characterization. *Z. Naturforsch.*, 43c, 850, 1988.

Durner, J., Gailus, V., and Böger, P., New aspects on inhibition of plant acetolactate synthase by chlorsulfuron and imazaquin. *Plant Physiol.*, 95, 1144, 1991.

Schloss, J. V., Interaction of the herbicide sulfometuron methyl with acetolactate synthase: a slow binding inhibitor, in *Flavins and Flavoproteins*, Bray, R. C., Engel, P. C. and Mayhew, S. G., Eds., Walter de Gruyter, Berlin, 1984, p. 737.

Westerfeld, W. W., A colorimetric determination of blood acetoin. *J. Biol. Chem.*, 161, 495, 1945.

CHAPTER 19

Ketol-Acid Reductoisomerase from Plants: Extraction and Enzymatic Assay

Jörg Durner, Petra Weiss, Oliver Knörzer, and Peter Böger

INTRODUCTION

The enzyme ketol-acid reductoisomerase (KARI, EC 1.1.1.86; also referred to as AHRI, acetohydroxy-acid reductoisomerase) is involved in the biosynthesis of the branched-chain amino acids valine, leucine, and isoleucine. The substrate, either 2-acetolactate or 2-aceto-2-hydroxybutyrate, is reduced by an NADPH-dependent reaction to 2,3-dihydroxy-3-methylbutyrate or 2,3-dihydroxy-3-methylvalerate, respectively. Previously, the enzyme has been purified to homogeneity from the stroma of spinach leaf chloroplasts (Dumas et al., 1989).

Recently, growth inhibition of plants and bacteria by the experimental compound Hoe 704 could be demonstrated (Schulz et al., 1988). The target enzyme KARI was found to be inhibited in a time-dependent and (with respect to the substrates) competitive manner. This contribution describes a fast and reliable procedure to extract and enrich KARI activity from higher plants.

0-87371-539-X/93/$0.00 + $.50

PROCEDURE

Cultivation

Basically, cultivation of plants was done as described for the ALS assay (Chapter 18). Since barley has been proven to be the species with the highest extractable ALS activity, it seemed to be a good source to obtain KARI. However, experiments with other species should be carried out in order to optimize the yield of active enzyme.

Equipment

Besides basic laboratory equipment, preparative (e.g., Sorvall RC-2) and microcentrifuges (e.g., Eppendorf) are required. Most importantly, a spectrophotometer equipped with a constant-temperature multiple-cell holder should be available. If not, a standard spectrophotometer and a water bath are an alternative, albeit a less comfortable one.

Herbicides

The herbicide used was the experimental compound Hoe 704, 2-methylphosphinoyl-2-hydroxyacetic acid.

Enzyme Extraction

The extraction was carried out at 4°C. Frozen shoots (100 g) were homogenized in a blender in 400 mL extraction buffer (50 m*M* tris(hydroxymethyl)amino methane (Tris)/HCl, pH 7.5, 5 m*M* $MgCl_2$, 1 m*M* EDTA, and 15% glycerol, v/v). Unbroken cells and cell debris were removed by a 7-min centrifugation at 20,000 × *g*. Subsequently, a fractionated ammonium sulfate precipitation was carried out. KARI precipitated between 35 and 60% saturation. Collection of the precipitates was done by centrifugation for 15 min at 27,000 × *g*. The final (60%) pellet was resuspended in extraction buffer (5 mL) and desalted on Sephadex G-25 PD-10 minicolumns equilibrated with extraction buffer containing 20% (v/v) glycerol. This preparation could be stored at 70 K for later use. The specific activity was between 0.36 and 0.53 μmol NADP per minute per milligram.

Sample Preparation

For further purification the extracted protein was desalted on PD-10 minicolumns equilibrated with potassium phosphate buffer (25 m*M*, pH 7.5) containing 4 m*M* $MgCl_2$ and 10% glycerol (v/v) and applied to a column packed with Red-Sepharose (same buffer). Elution was performed by a KCl gradient (0 to 1 *M*). Fractions containing KARI activity (0.3 *M* KCl) were pooled and stored at −20°C until later use. The specific activity of this preparation was between 1.1 and 1.5 μmol NADP per minute per milligram.

Substrate Synthesis

Neither of the two substrates of KARI is commercially available. Therefore, 2-acetolactate was produced by enzymatic hydrolysis of ethyl-2-acetoxy-2-methyl-acetoacetate (available from Aldrich Chemicals, Steinheim, Germany) by an esterase from hog liver (purchased from Boehringer, Mannheim, Germany). The reaction mixture was: 0.9 mL esterase (stock, 10 mg/mL with a specific activity of 130 U/mg), 0.15 mL ester (final concentration of 0.1 *M*), and 6.45 mL Tris/HCl, pH 8 (final volume: 7.5 mL). Hydrolysis of the ester was carried out for 15 min at 24°C. To maintain a constant pH of 8, continuous addition of 1 *N* NaOH is necessary (the pH should never fall below pH 8; the product of the reaction, 2-acetolactate, is very unstable below this value). The yield of 2-acetolactate was 30% (30 m*M*). The solution was stored in aliquots at −20°C.

Enzyme Assay

The assay mixture contained 50 m*M* Tris/HCl, pH 7.5, 4 m*M* $MgCl_2$, 0.2 m*M* NADPH, 0.3 m*M* 2-acetolactate, plus enzyme in a final volume of 1 mL. Protein in the assay was between 1 and 2 mg. The temperature was 30°C and the reaction time 10 min. For inhibition experiments, the inhibitor was dissolved in 50 m*M* Tris/HCl, pH 8, and stored as a 5 m*M* stock at 4°C. The enzyme reaction was detected as NADPH oxidation at 340 nm.

RESULTS

Detection of KARI Activity

For the present the most puzzling question is which grade of purification of KARI is suitable to determine either biochemical parameters as substrate affinity or inhibition by new compounds. By no means could any KARI activity be detected in crude extracts. The use of the preparation obtained by fractionated ammonium sulfate precipitation led to a K_m for acetolactate of 2.5 m*M* (data not shown).

In contrast, with partially purified enzyme (Red-Sepharose fraction) the K_m has been determined as 82 μ*M* (Figure 1). A strict Michaelis-Menten kinetics with respect to acetolactate was observed. Substrate saturation was between 0.4 and 0.6 m*M*.

Inhibition of KARI

Inhibition experiments were carried out with the partially purified KARI preparation (Red-Sepharose fraction). Similar to inhibition of ALS by sulfonylureas or imidazolinones, the extent of inhibition of KARI by Hoe 704 increased with time until a final steady state had been attained. The time of transition from initial to steady-state inhibition was between 20 and 40 min

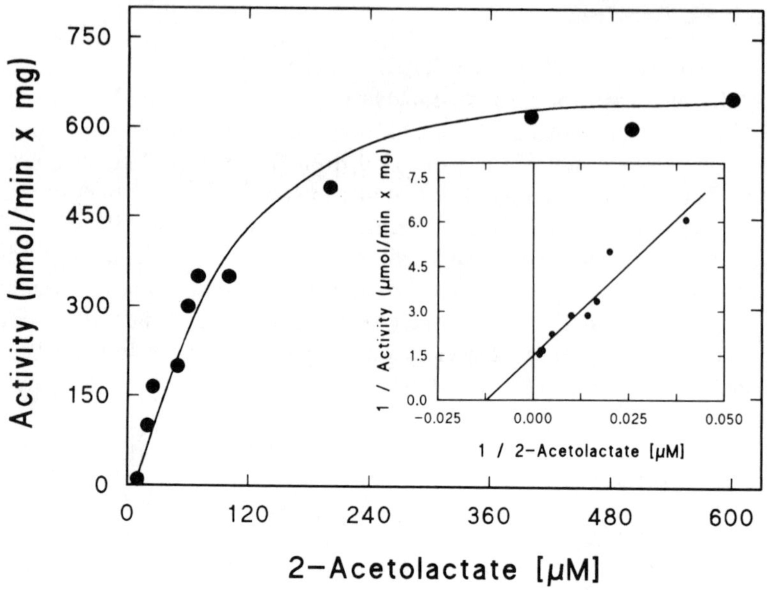

Figure 1. Substrate saturation curve for 2-acetolactate with KARI from etiolated barley shoots. The KARI assay was conducted under the standard assay conditions as described, except that the concentration of 2-acetolactate was varied as indicated. The inset is a double-reciprocal plot of the data.

(data not shown). Therefore, the fixed-time assay of 10 min allowed the determine of an apparent I_{50} value (0.14 μM; Figure 2), but not an exact characterization of the mode of action of Hoe 704.

In contrast to the partially purified enzyme in crude extracts or preparations applied to ammonium sulfate precipitation, no inhibition by Hoe 704 could be observed.

COMMENTS

Without a doubt, the procedure presented to screen inhibitors of KARI is not as simple as that for ALS inhibitors (see Chapter 18). First, the substrates of KARI (2-acetolactate or 2-aceto-2-hydroxybutyrate) are not available commercially: second, in crude extracts from plants no KARI and, therefore, of course, no inhibition can be detected. However, plant cultivation and extraction are quite easy; both the synthesized 2-acetolactate and the partially purified enzyme can be stored frozen. Therefore, there is every reason to scale up the presented substrate synthesis and enzyme preparation. Besides this, characterization of the purified enzyme and further research on the mode of action of KARI inhibitors are urgently needed.

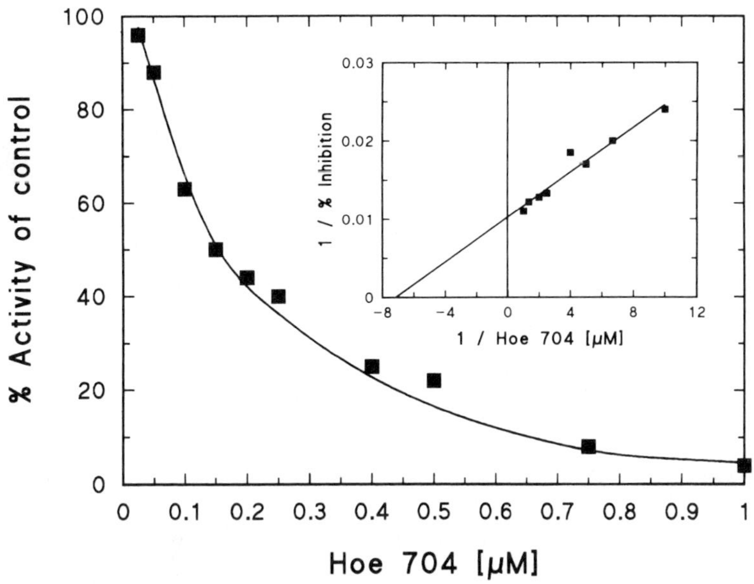

Figure 2. Dose response for the inhibition of KARI from etiolated barley shoots by Hoe 704. The values are given as the percent of an untreated control. KARI assay was conducted under the standard assay conditions as described in the text. The specific activity of the control was 1.35 μmol NADPH per minute per milligram. The inset is a double-reciprocal plot of percent inhibition vs. inhibitor concentration.

ACKNOWLEDGMENT

This study was supported by the Fonds des Chemischen Industrie.

REFERENCES

Dumas, R., Joyard, J. and Douce, R. Purification and characterization of acetohydroxyacid reductoisomerase from spinach chloroplasts. *Biochem. J.*, 262, 971, 1989.

Schulz, A., Spönemann, P., Köcher, H. and Wengenmayer, F., The herbicidally active experimental compound Hoe 704 is a potent inhibitor of the enzyme acetolactate reductoisomerase. *FEBS Lett.*, 238, 375, 1988.

CHAPTER 20

Screening of Inhibitors (Antimetabolites) of the Biosynthesis or Function of Amino Acids or Vitamins with a *Lemna* Assay

Reynold Chollet

INTRODUCTION

Biosynthetic pathways specific to plants (and occasionally to microorganisms) are privileged targets for herbicides (especially when toxicological aspects are considered).

Photosynthesis was the first one to be exploited some 40 years ago. In recent years, however — more precisely, after the discovery of glyphosate (an inhibitor of the biosynthesis of aromatic amino acids) and especially of the sulfonylurea and imidazolinone herbicides (inhibitors of the biosynthesis of branched-chain amino acids) — the amino acid biosynthesis pathways have attracted the attention of most (if not all) agrochemical companies (see Powell and Rees, 1989).

Similarly, in this author's view, vitamin biosynthesis pathways also constitute good potential targets for new herbicides (e.g., asulam inhibits the biosynthesis of folic acid).

A very simple, inexpensive, short, selective, and reproducible bioassay using *Lemna minor* allows the screening of compounds that inhibit the biosynthesis of specific amino acids or vitamins (see Comment 1).

0-87371-539-X/93/$0.00 + $.50

Principle of the Assay

The growth inhibition caused by an inhibitor of a specific amino acid or vitamin is strongly reduced or even totally alleviated in the presence of that amino acid or vitamin.

PROCEDURE

Materials

Materials required include petri dishes (diameter 20 cm) and microscope lamella boxes (22 × 22 × 18 mm) with perforated lids (one hole, diameter 2 mm; see Comment 2).

Inhibitors

Inhibitors used were asulam: methyl((4-aminophenyl)sulfonyl)carbamate; sulfamoxazole: 4-amino-*N*-(4,5-dimethyl-2-oxazolyl)benzenesulfonamide; azetindine-2-carboxylic acid (A2C); thiazolidine-4-carboxylic acid (T4C); *p*-F-phenylalanine (F-PHE); L-mimosine: α-amino-3-hydroxy-4-oxo-1(4H)-pyridinepropanoic acid (MIM); 6-F-tryptophan (F-TRP); and 1,2,4-triazole-3-alanine (T3A).

Solvents

The following solvents can be used if their final concentration is not higher than 1% (v/v): acetone, isopropanol, xylene, dimethylsulfoxide, and tetrahydrofuran. Ethanol and *N,N*-dimethylformamide should be kept below 0.5%. Methanol and *N*-methyl-2-pyrrolidone should be kept below 0.1%, and acetonitrile should be avoided.

Cultivation

Lemna minor (harvested in a pond and cloned) is grown in mineral medium consisting of KNO_3, 202 mg/L; KH_2PO_4, 50.3 mg/L; K_2HPO_4, 27.8 mg/L; K_2SO_4, 17.4 mg/L; $MgSO_4 \times 7\ H_2O$, 49.6 mg/L; $CaCl_2$, 11.1 mg/L; Na_2-EDTA, 10 mg/L; $FeSO_4 \times 7\ H_2O$, 6 mg/L; H_3BO_3, 5.72 mg/L; $MnCl_2 \times 4\ H_2O$, 2.82 mg/L; $ZnCl_2$, 0.6 mg/L; $Na_2MoO_4 \times 2\ H_2O$, 0.059 mg/L; $CuCl_2 \times 2\ H_2O$, 0.078 mg/L; $CoCl_2 \times 6\ H_2O$, 0.054 mg/L; the pH is adjusted to 6.5.

A nursery culture is maintained in petri dishes (400 mL of mineral medium each) placed in a growth chamber at 20°C and under continuous light (6000 to 8000 lux). The fronds are washed twice weekly with 0.01 *M* NaClO (maximum 40 to 60 sec: see Comment 3), then rinsed thoroughly with tap water before transferring them into fresh medium. At the same time the frond population is reduced in order to cover about 10 to 20% of the petri dish surface, and the bad-looking, senescing fronds are discarded.

Treatments (Primary Assay)

A block **N** (= normal) of lamella boxes (see Materials) is filled with 3 mL of mineral medium; then healthy pairs of *Lemna* are selected (see Comment 4), and one pair is placed in each box.

A second block **M** (= metabolites) is similarly prepared (medium + *Lemna*), except that the growth medium additionally contains selected metabolites (amino acids and/or vitamins) at the highest admissible concentration (see Comment 5).

The screening compounds are then pipeted at the desired concentration (see Comment 5) into the boxes of the two blocks, and the perforated lids are put on. The two blocks **N** and **M** are then placed in growth chambers at 25°C and under light conditions (16-hr photoperiod, 8000 to 10,000 lux).

Rating

The rating occurs visually by counting the number of new fronds after 6 days of incubation (see Comment 6).

In the case of a significant protection in block **M** for a compound that is inhibitory in block **N**, a fine tuning assay is run.

Fine Tuning Assay (Secondary Assay)

Each compound for which a protection has previously been observed in the presence of the metabolite mixture (block **M**) is retested alone and in the presence of each isolated metabolite (amino acid or vitamin).

A protective effect by only one of them indicates which biosynthesis pathway is affected, and a dose-response assay is run. When several of the metabolites are protective by themselves, the site of activity of the inhibitor is generally undetermined (see Comment 7).

Dose-Response Assay (Tertiary Assay)

The inhibitor is retested at several concentrations in the absence and in the presence of its specific protectant.

RESULTS

Table 1 shows a protocol for an experiment with some specific inhibitors ± selected metabolites (0.1 m*M* folic acid = FOL, 0.1 m*M* thiamine = THI, 0.1 m*M* biotin = BIO, 0.3 m*M* L-proline = PRO, 0.1 m*M* L-histidine = HIS, 0.3 m*M* L-phenylalanine = PHE, and 0.3 m*M* L-tryptophan = TRP) given as a mixture (primary assay) or separately (secondary assay = fine tuning) (see Comment 8). The metabolite mixture protects against each of the inhibitors tested (primary assay), and in each case the protection can be assigned to only one of the metabolites present in the mixture (secondary

Table 1. Protocol for the Screening of Specific Antimetabolites with *Lemna minor*

Inhibitor[a] (50 μ*M*)	Protective Effect of: Metabolite Mixture (Primary Assay)	Specific Metabolites[b] Alone (Secondary Assay = Fine Tuning)						
		FOL	THI	BIO	PRO	HIS	PHE	TRP
Asulam	Yes	Yes	No	No	No	No	No	No
Sulfametazole	Yes	Yes	No	No	No	No	No	No
A2C	Yes	No	No	No	Yes	No	No	No
T4C	Yes	No	No	No	Yes	No	No	No
T3A	Yes	No	No	No	No	Yes	No	No
F-PHE	Yes	No	No	No	No	No	Yes	No
MIM	Yes	No	No	No	No	No	Yes	No
F-TRP	Yes	No	No	No	No	No	No	Yes

[a] See text for complete inhibitor names.
[b] Metabolite concentration: folic acid (FOL), thiamine (THI), biotin (BIO), L-histidine (HIS) = 100 μ*M*; L-proline (PRO), L-phenylalanine (PHE), L-tryptophan (TRP) = 300 μ*M*.

assay). Figure 1 shows dose-response curves (tertiary assay) for the inhibitors from Table 1 (± their specific protectant).

The herbicide asulam (Figure 1A) and the antibacterial agent sulfamoxazole (Figure 1B) are both antimetabolites of folic acid. They are known to inhibit the enzyme dihydropteroate synthase (competitively to *p*-aminobenzoate). Azetidine-2-carboxylate (A2C: Figure 1C) and thiazolidine-4-carboxylate (T4C: Figure 1D) are both antimetabolites of proline, A2C can replace proline in newly synthesized proteins, and T4C is a known inhibitor of proline biosynthesis.

p-F-phenylalanine (F-PHE: Figure 1E) and L-mimosine (Figure 1F) are both known antimetabolites of phenylalanine, and 6-F-tryptophan (F-TRP: Figure 1G) and 1,2,4-triazole-3-alanine (T3A: Figure 1H) are known antagonists of tryptophan and histidine, respectively.

COMMENTS

1. Theoretically, the present assay should also allow in some cases the detection of compounds that interfere with the function of a specific metabolite (e.g., incorporation of amino acids into proteins, or cofactor role of vitamins).
2. Multiple-well plates (e.g., six rows of four holes) can be used in place of single boxes; however, there is the risk of a contamination of several wells by a single highly volatile compound. The lid perforation favors gas exchange and slows down the natural senescence.
3. There is no need for absolute sterile conditions. However, the cultures are often contaminated by undetermined algae that can be maintained at an acceptable level by the NaClO treatment. Higher NaClO concentrations or longer NaClO treatment could be phytotoxic for *Lemna minor*.

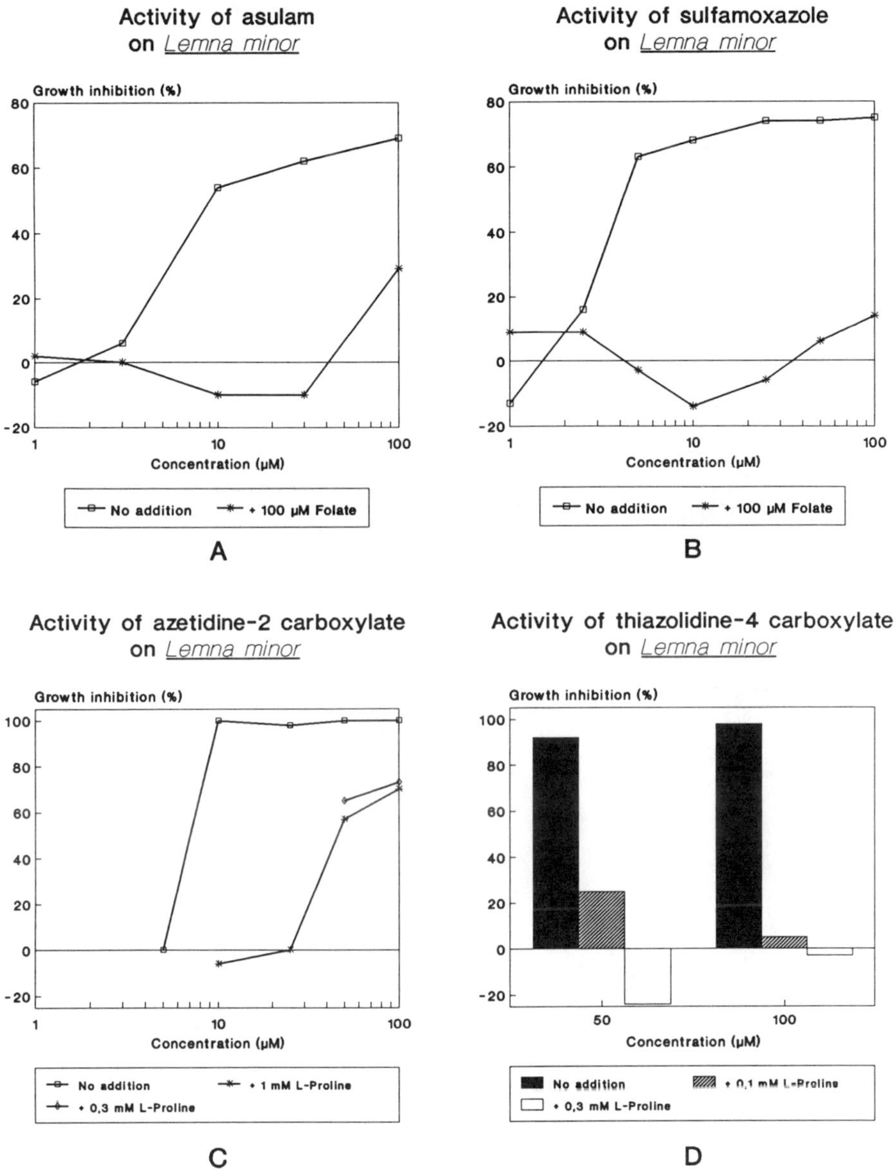

Figure 1. (**A,B,C,D,E,F,G,H**). Activity of several antimetabolites on *Lemna minor* (in the absence or in the presence of their respective protectant).

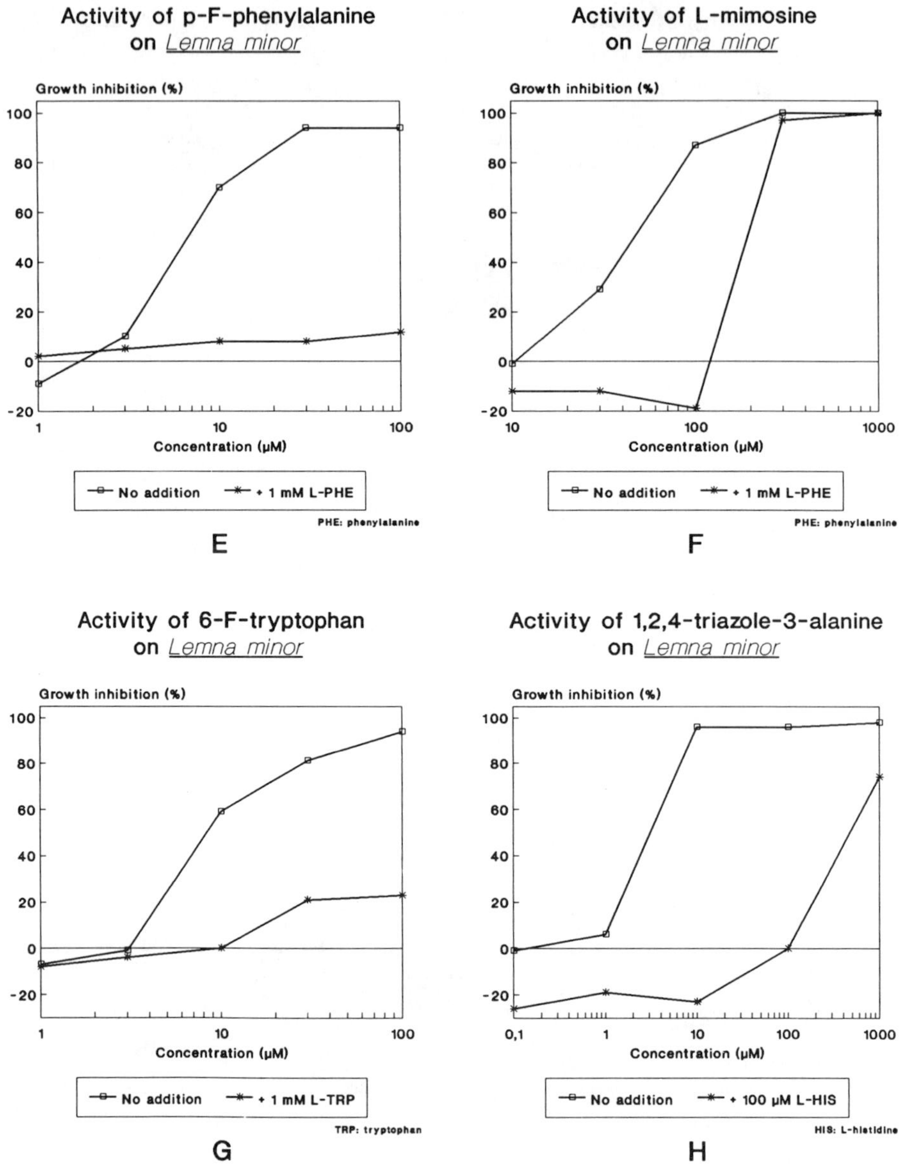

Figure 1 (continued).

4. A *Lemna minor* population includes older (physiologically mature) and younger (immature) fronds. The older and younger fronds may show different sensitivity to herbicides. A frond pair (two fronds still linked together) always consists of an older and a younger frond; thus, the selection of pairs provides a more homogeneous starting material (if the physiological stage and the herbicide sensitivity are considered) than a selection of single fronds, in which case a treatment of two fronds could concern two older or two younger fronds in addition to the previous case (one older + one younger). Further, the separation of pairs from larger groups is much easier than the separation of single fronds (risk of injury).
5. Each metabolite (amino acid or vitamin) is added at the highest concentration that is not inhibitory to *Lemna* in the mixture (synergistic inhibitions are possible, e.g., with threonine + lysine). In general, the metabolites are added between 0.05 and 1 m*M* and the screening compounds are tested at 50 μ*M*.
6. Only the newly formed fronds are counted, according to their size: the fully developed fronds as single units, the smaller ones as one half or one fourth of a unit. This mode of counting better quantifies the real biomass growth, especially when a normal number of very small fronds is formed. An incubation of 6 days is preferable for the final counting, but generally a visual assessment is already possible after 4 days.
7. A multiple protection generally does not allow one to determine which pathway is inhibited, except when the protective metabolites belong to the same pathway; e.g, protection by phenylalanine and by tryptophan would indicate an inhibition of the shikimate pathway. However, glyphosate cannot be used as the standard for that pathway because it is not inhibitory to *Lemna* (inactivation of the herbicide occurs through its chelation by the calcium present in the growth medium).
8. The list of metabolites given here is just an example and can be changed or completed as desired after possible synergistic inhibitions have been ruled out (see Comment 5). Metabolites that are inhibitory by themselves above 0.05 m*M* should be avoided (e.g., valine and leucine).

REFERENCE

Powell, K. and Rees, S. B., Screening techniques for amino-acid biosynthesis inhibitors, in *Prospects for Amino Acid Biosynthesis Inhibitors in Crop Protection and Pharmaceutical Chemistry,* Copping, L. G., Dalziel, J., and Dodge, R. D., Eds., Monograph No. 42, British Crop Protection Council, Farnham, Surrey, U.K., 1989, p. 31.

Fatty Acid Formation

CHAPTER 21

Assay of Plant Acetyl-CoA Carboxylase and Its Specific Inhibition by Some Graminicides

Hartmut K. Lichtenthaler and Manfred Focke

INTRODUCTION

De novo fatty acid biosynthesis in higher plants proceeds in chloroplasts and other plastid forms (etioplasts, leucoplasts) and comprises the enzymes acetyl CoA synthetase (ACS). acetyl-CoA carboxylase (ACC), and the fatty acid synthetase complex (FAS) and yields predominantly C 16:0 and 18:1 fatty acids. In recent years two groups of graminicides, the aryloxyphenoxypropionic acids (e.g., diclofop) and the cyclohexane-1,3-diones (e.g., cycloxydim, sethoxydim), were developed which specifically inhibit the ACC of members of the Poaeceae, but not the ACC from other plant families such as dicotyledons or monocotyledons outside the Poaceae (Lichtenthaler, 1989; Lichtenthaler et al., 1989).

There are two possible ways to investigate the inhibitory capacity of an unknown compound on the ACC of plants.

1. The first possibility is to apply isolated intact plastids and to measure the incorporation of radioactive acetate into the total fatty acid fraction as a marker for the activity of ACC. This method has the advantage of its relative simplicity and the fact that isolated plastids reflect semi-*in vivo* conditions. This method also allows the determination of reliable I_{50}

0-87371-539-X/93/$0.00 + $.50

values. One disadvantage may be that the tested xenobiotics may inhibit not only the ACC, but also other enzymes involved in the complete fatty acid biosynthesis pathway (e.g., ACS or FAS). On the other hand, this assay system offers the possibility to study the effect of inhibitors of the FAS such as the antibiotics cerulenin or thiolactomycin (Feld et al., 1989).

2. The second possible method to investigate inhibition effects on the ACC is to isolate and purify the plant ACC to such an extent that one can achieve reproducible enzyme assays. This second possibility is much more expensive and time consuming than the first one. It will, however, be applied when a synthetic compound has shown an inhibitory effect on the *de novo* fatty acid biosynthesis at the level of whole plastids. It should also be noted in this respect that the I_{50} values obtained with a partially purified ACC fraction tend to be two to five times higher than those of the quasi-*in vivo* assay system of isolated chloroplasts or etioplasts.

The chloroplast test system for ACC inhibitors is very efficient and reproducible since the cofactors ATP and NADPH, which are required for fatty acid biosynthesis, are regenerated by the photosynthetic light reactions. It is a reliable test system for the assay of all compounds which do not directly affect the photosynthetic light reactions. Compounds which block the photosynthetic ATP and NADPH formation — e.g., inhibitors of the photosynthetic electron transport such as diuron or bentazon — decrease the rate of fatty acid biosynthesis in isolated chloroplasts by decreasing the level of photosynthetically formed ATP and NADPH. This problem may be overcome by using isolated etioplasts, which exhibit *in situ* ATP And NADPH synthesis or can accept exogenously offered ATP and NADPH for fatty acid biosynthesis (see Kobek and Lichtenthaler, 1989).

PROCEDURE

Cultivation

It is a general rule that young plants or young parts of plants (e.g, meristematic tissue, young leaves) are the best source for investigating *de novo* fatty acid biosynthesis. These exhibit higher ACC activities than older leaves or plant tissues. Plants are cultivated on peat enriched with nutrients, at room temperature and a relative humidity of approximately 50 to 70%. At harvest the plants are about 1 week old or somewhat less, except for spinach (2 to 3 weeks). The dicot plants pea and spinach and the monocot plants oat, barley, and maize are good test plants for the isolation of chloroplasts and etioplasts as well as the ACC enzyme. The use of other plants has to be checked empirically. Although the dicot plants pea and spinach are not sensitive toward the ACC inhibitors of the cyclohexanedione or aryloxyphenoxypropionic acid type, they are applied for comparative and control reasons.

Equipment

Measurement of the ACC activity either with plastids (chloroplasts, etioplasts) or in a direct ACC assay requires good biochemical equipment, including an isotopic laboratory. A protein high-pressure liquid chromatography (HPLC) system with anion exchange chromatography is an advantage for purifying ACC, but also can be substituted by a more time-consuming classical chromatography column system.

Herbicides

The routinely used herbicides are: cycloxydim, 2-(1-ethoxyimino)butyl-5-(thian-3-yl)-2-cyclohexane-1,3-dione; sethoxydim, 2-(1-ethoxyimino)butyl-5-(2-ethylthiopropyl)-3-hydroxy-2-cyclohexen-1-one; and diclofop, 2-(2,4 dichlorophenoxy)phenoxy)propionic acid.

The xenobiotics may be dissolved in 100% methanol and diluted with slightly alkaline buffer to prevent self-acidolysis (cyclohexane-1,3-diones and aryloxyphenoxypropionic acids are both weak acids). However,the end concentration of methanol in the assay should not exceed 0.2%. The assay mixture of the controls also contains 0.2% methanol. In the case of the aryloxyphenoxypropionic acid-type herbicides, the free acid form, which is the true active substance, should be used and not the ester form, which is routinely applied in field experiments.

Isolation of Plastids

To isolate highly active intact plastids, one must work quickly and at low temperature. Plant material (approximately 20 to 50 g fresh weight green leaves or 150 to 200 g fresh weight etiolated leaves) should be homogenized with short strokes of a homogenizer with very sharp blades (e.g., replaceable razor blades) in approximately 150 mL standard homogenization buffer. The latter consists of 500 m*M* sorbitol, 50 m*M* potassium phosphate (pH 8), and 2 m*M* magnesium chloride. The homogenate should be filtered through several layers of cheesecloth and centrifuged for 90 sec at 2000 $\times$ *g* in the case of **chloroplasts** or for 90 sec at 4000 $\times$ *g* in the case of **etioplasts**. In special cases one may use in the homogenization buffer Tricine (pH 9; 100 m*M*) instead of potassium phosphate. To compare different plastid isolations, the content of chlorophylls (a + b) and carotenoids (x + c) of the plastid suspension are determined in 100% methanol (Lichtenthaler, 1987). In the case of chloroplasts, approximately 15 μL suspension are dissolved in 1.5 mL methanol, and in the case of etioplasts 150 μL suspension are dissolved in approximately 1.5 mL methanol.

Plastid Assay System

The plastidic pellet has to be resuspended immediately (50 to 100 μg chlorophyll per milliliter in the case of chloroplasts and 10 to 30 μg carotenoids per milliliter in the case of etioplasts) and incubated at room temperature with 300 m*M* sorbitol, 50 m*M* potassium phosphate (pH 7.9 for chloroplasts, pH 8.1 for etioplasts), 30 m*M* sodium hydrogen carbonate, 2.5 m*M* dithioerythritol (DTE), 4 m*M* ATP, 0.5 m*M* coenzyme A, and 0.5 m*M* magnesium chloride in a total volume of 0.5 mL including 1 μCi ^{14}C-acetate (= 37 kBq). The reaction is started by adding the plastid suspension of approximately 100 μL. In the case of chloroplasts the suspension medium, which is shaken at room temperature, has to be illuminated either from the bottom or the top (approximately 1400 μmol m^{-1} sec^{-1} photosynthetically active radiation) and is protected against heating using an infrared filter (e.g., a water cuvette). The etioplast suspension medium has to be incubated in the dark. After 20 min the reaction is stopped by adding 0.5 mL 30% potassium hydroxide, and the glycerolipids are saponified for 90 min at 80°C. After acidification with 0.25 mL 6 *M* sulfuric acid and 0.25 mL 30% trichloroacetic acid, the fatty acids are extracted twice with 2.5 mL and 1.5 mL light petrol (b.p. 50 to 70°C). The latter is placed in a scintillation flask. After evaporation of the light petrol and addition of scintillation cocktail (shaking overnight) the radioactivity is measured.

The biosynthetic capacity for fatty acids of the control chloroplast suspensions should lie in the range of 30 to 150 nmol acetate per milligram chlorophyll per hour, and in the case of etioplast suspensions it should lie in the range of 300 to 600 nmol acetate per milligram carotenoids per hour. For the 0.5-mL reaction mixture given above, this is equivalent to 3000 Bq in the case of chloroplasts (with 10 μg chlorophyll per sample and a 20-min incubation time) or 300 Bq in the case of etioplasts (with 3 μg carotenoids and a 20-min incubation time). On a protein basis, the acetate incorporation in chloroplasts and etioplasts may lie in the range of about 1 to 3 nmol acetate per milligram protein per hour. Chromatographic tests revealed that more than 98% of the radioactivity found in the petrol phase is incorporated into fatty acids.

ACC Isolation

A total of 200 g leaves, including shoots of young seedlings (e.g., maize, barley, pea), are homogenized in the presence of insoluble polyvinylpolypyrrolidone in 400 mL buffer containing 100 m*M* Tricine (pH 8) and protease inhibitors (2 m*M* aminocaproic acid, 2 m*M* benzamidine, and 0.2 m*M* phenylmethylsulfonylfluoride). After addition of glycerol to a final concentration of 10%, the homogenate is centrifuged for 15 min at 10,000 × *g*. Saturated ammonium sulfate solution is added to the supernatant to a final saturation concentration of 40%. After centrifugation the ACC-containing pellet is resuspended in 10 m*M* Tricine (pH 8), 1% glycerol, 50 m*M* potassium chloride, 0.05% Tween 20, and protease inhibitors (see above) and then desalted with the help of small desalting columns (e.g., Sephadex G-25, PD-10) using the same buffer. The desalted protein solution may be used for inhibitor studies, but it is preferable to include a purification step with anion exchange chromatography (e.g, Fractogel™ EMD TMAE-650 (M), Merck, Germany). In our HPLC separation system the ACC enzyme fraction is eluted between 300 and 400 m*M* potassium chloride (linear gradient from 0 to 1 *M* potassium chloride in 1 hr). ACC may be stored frozen at −80°C for 2 weeks.

ACC Assay

The principle of this assay is to measure the heat- and acid-stable incorporation of ^{14}C-hydrogen carbonate into acetyl-CoA (= formation of malonyl-CoA). Included in a total volume of 0.05 mL are: 100 m*M* n-tris(hydroxymethyl)methylglycine (Tricine)-NaOH, pH 8, 2.5 m*M* magnesium chloride, 1 m*M* ATP, 1 m*M* DTE, 0.3 m*M* acetyl-CoA, 20 m*M* ^{14}C-hydrogen carbonate (= 3700 Bq), and approximately 50 μg of protein from the post-anion-exchange chromatography fractions. The incubation buffer has to be degassed and stored under ligroine to prevent dilution of the radioactivity with atmospheric carbon dioxide. After 10 min at 30°C, the reaction is stopped by addition of 25 μL 6 *N* hydrogen chloride and the liquid is transferred to a scintillation tube. After evaporation of the liquid to dryness at 90°C in an oven, the remaining radioactivity is measured in a liquid scintillation counter. Controls without acetyl-CoA have to be included in each series in order to subtract the background. Samples without inhibitors exhibit up to 40 Bq within the 10-min incorporation time applied.

RESULTS

Examples of the inhibition of the plant fatty acid biosynthesis and ACC activity by the graminicides cycloxydim and diclofop using the two test systems, isolated plastids and purified ACC, are presented in Tables 1 and 2. The I_{50} values of a second cyclohexanedione derivative, sethoxydim, which is also applied in the comparative routine analysis are slightly higher than those shown here for cycloxydim. The basic procedure used to determine I_{50} values, which is valid for sigmoidal inhibition curves, is shown in Figure 1. A good linear correlation (Figure 1B) is a measure of the accuracy of the determination of the I_{50} values. At linear correlation values below 0.9 the I_{50} values are not reliable. The results indicate that the strength of the inhibition depends on the species tested and the inhibitor used. If in the plastid test a potent inhibitor of *de novo* fatty acid biosynthesis is found, one has to check using the direct ACC assay whether this inhibitor is indeed an ACC inhibitor. *Vice versa,* if an active ingredient is a strong inhibitor in the ACC assay it should also exhibit similar effects at the level of isolated plastids. Although

Table 1. I_{50} Values of the Graminicides Cycloxydim and Diclofop in the Testing of Isolated Plastids of Plants with Different Taxonomic Position and of Various Poaceae

Species	I_{50} (μM)	
	Cycloxydim	**Diclofop**
Chloroplasts		
Chlorophytum comosum	>500	ca. 280
Nicotiana tabacum (tobacco)	>500	ca. 250
Pisum sativum (pea)	>500	200
Spinacia oleracea (spinach)	>500	150
Poaceae		
Festuca ovina	100	0.4
Festuca rubra	100	0.4
Poa annua	10	0.2
Poa pratensis	0.2	0.1
Avena sativa (oat)	0.15	0.1
Festuca arundinacea	0.1	0.02
Etioplasts		
Pisum sativum (pea)	>500	135
Avena sativa (oat)	0.3	0.3

in most cases similar inhibition effects are found on the level of isolated plastids and the purified ACC enzyme, there may exist differences in some cases where the plastid envelope might function as a certain diffusion barrier for the compound to be tested.

COMMENTS

With the two test systems described above it is possible to screen a great number of potential ACC inhibitors in a relatively short time and to determine their I_{50} values. At first this will be done with the more rapid and easy to handle plastid test system. On a second level, potential inhibiting compounds (e.g., those with low I_{50} values) will be confirmed with the ACC assay to

Table 2. Example of the Inhibition of the Acetyl-CoA Carboxylase (ACC) from Maize Leaves Purified with Anion Exchange HPLC by the Herbicides Cycloxydim and Diclofop.

Concentration of Inhibitor (*M*)	Cycloxydim	Diclofop
$5 \cdot 10^{-8}$	—	107
10^{-7}	86	104
$5 \cdot 10^{-7}$	—	58
10^{-6}	58	42
$5 \cdot 10^{-6}$	29	14
10^{-5}	19	10

Note: —, indicates no data.

[a] Values are in percent of control. the I_{50} values measured for maize ACC are approximately 1.5 μM (cycloxydim) and approximately 0.7 μM (diclofop).

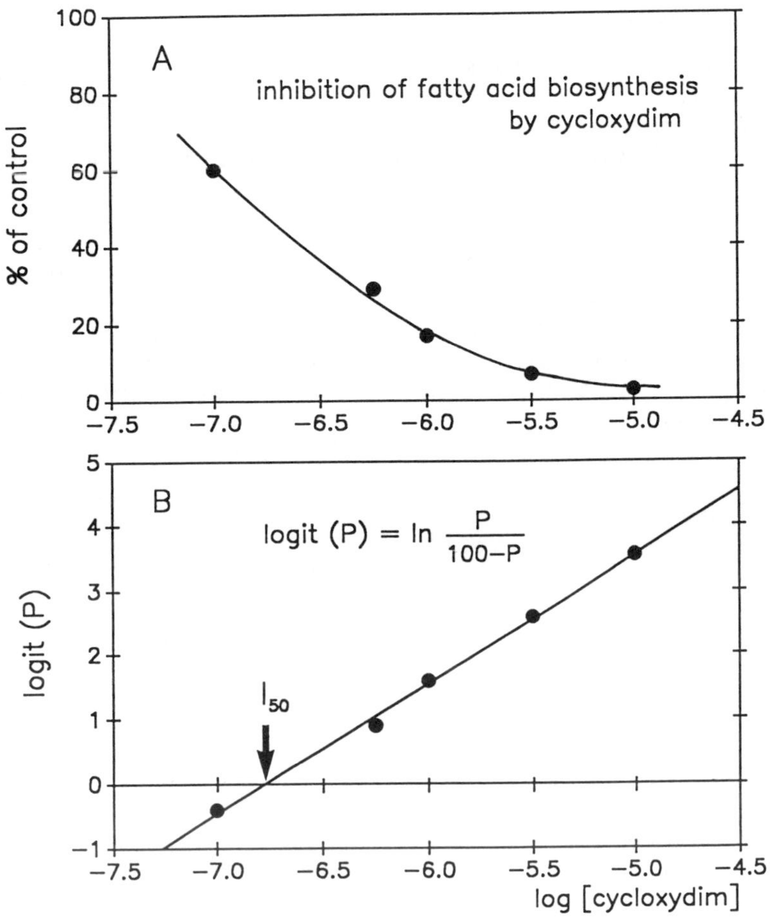

Figure 1. Graphic representation of the inhibition of *de novo* fatty acid biosynthesis of isolated oat chloroplasts by cycloxydim. (A) Inhibition of ^{14}C-acetate incorporation into the total fatty acid fraction. (B) Linearization of the inhibition plot of (A) by the logit transformation using the equation indicated (P = percentage inhibition at different herbicide concentrations). The I_{50} value is obtained at logit (P) = 0. The I_{50} value in this case is 0.14 μM. Linear correlation in (B): r − 0.996.

check further whether these compounds are true ACC inhibitors. In addition, more detailed kinetic studies can be performed in the direct ACC enzyme assay in order to determine the type of inhibition and interference with the ACC enzyme.

REFERENCES

Feld, A., Kobek, K., and Lichtenthaler, H. K., Inhibition of de novo fatty acid biosynthesis in isolated chloroplasts by different antibiotics and herbicides. *Z. Naturforsch.,* 44c, 976, 1989.

Kobek, K. and Lichtenthaler, H. K., Inhibition of de novo fatty acid biosynthesis in isolated etioplasts by herbicides. *Z. Naturforsch.,* 44c, 669, 1989.

Lichtententhaler, H. K., Chlorophylls and carotenoids: pigments of photosynthetic biomembranes. *Methods Enzymol.,* 148, 350, 1987.

Lichtenthaler, H. K., Mode of action of herbicides affecting acetyl-CoA carboxylase and fatty acid biosynthesis. *Z. Naturforsch.,* 45c, 521, 1989.

Lichtenthaler, H. K., Kobek, K., and Focke, M., Differences in sensitivity and tolerance of monocotyledonous and dicotyledonous plants towards inhibitors of acetyl-Coenzyme A carboxylase. Brighton Crop Protection Conference — Weeds, 4D-9, 173, 1989.

CHAPTER 22

Activity and Inhibition of Acetyl-CoA Carboxylase Determined by HPLC

Gerhard Paul

INTRODUCTION

The cyclohexane-1,3-diones (e.g., cycloxydim, sethoxydim, tralkoxydim) and aryloxyphenoxypropionates (e.g., diclofop, haloxyfop, quizalafop) are selective herbicides for postemergence control of grasses in dicotyledonous crop plants. Although they are representatives of two distinct chemical classes, both are *in vitro* inhibitors of acetyl-CoA carboxylase (ACCase) in gramineous species (Burton et al., 1987, 1989; Focke and Lichtenthaler, 1987). Usually the ACCase *in vitro* reaction is measured by incorporating ^{14}C-carbonate into an acid-stable fraction without qualitative analysis of the resulting products. This chapter describes a nonradioactive high-pressure liquid chromatography (HPLC) method for simultaneous quantitation of aceytl- and malonyl-CoA on RP18 columns. This method, which is based on a procedure published in 1981 (Corkey et al., 1981), was developed in order to quickly analyze these two derivatives and free CoA. It can also be used to determine I_{50} values of ACCase-inhibitors.

0-87371-539-X/93/$0.00 + $.50

PROCEDURE

Plant Material

Corn (*Zea mays*) was germinated in darkness in vermiculite in an incubation chamber at 26°C. After 4 days the seedlings were transferred to constant light and room temperature, and they were harvested on day 7 after planting. The leaves were either used immediately for the preparation of the enzyme or stored at −80°C for up to 6 months without significant loss of ACCase activity.

Preparation of ACCase

Plant material (200 g) was homogenized in 400 mL buffer at 4°C (100 m*M* tris(hydroxymethyl)aminomethane (Tris)/HCl, pH 8.0, 20% glycerol, 14 m*M* mercaptoethanol, 15 m*M* $NaHCO_3$, 2 m*M* EDTA, and 0.4 m*M* phenylmethylsulfonylfluoride (PMSF)), filtered through eight layers of cheesecloth, and incubated with 55 g Dowex 1 × 2 for 15 min at 4°C. After a second filtration through cheesecloth, further contaminants were removed by precipitation with Polymin P (high-moleular-weight polyethylenimin, BASF; final concentration 0.01%) and centrifugation at 20,000 × *g* (see Egin-Bühler et al., 1980). The ACCase activity was collected by adding solid $(NH_4)_2SO_4$ to the supernatant up to a final concentration of 40% saturation. After centrifugation at 20,000 × *g*, the pellet could be stored for several weeks or used for enzyme assays after desalting on a Sephadex G-25 column (Pharmacia; 100 m*M* Tris/HCl, pH 8.0, 20% glycerol, 0.4 m*M* PMSF).

Assay for ACCase Activity

Varying amounts of ACCase and inhibitors were incubated in a total volume of 200 μL 100 m*M* Tris/HCl, pH 8.0, 1 m*M* ATP, 2.5 m*M* $MgCl_2$, 50 m*M* KCl, 15 m*M* $NaHCO_3$, and 50 μg acetyl-CoA. The reaction was started by adding acetyl-CoA and stopped after 5 min incubation at 30°C by adding 50 μL 9% perchloric acid. The precipitated protein was removed by centrifugation. A 200 μL volume of the supernatant was neutralized by adding 60 μL KOH/K_2HPO_4 (1 *M* KOH in 0.5 *M* K_2HPO_4), the precipitated $KClO_4$ was centrifuged, and 100 μL of the supernatant were analyzed by HPLC.

Equipment

The HPLC system consisted of a Waters 600E multisolvent delivery system (mixing of solvents on the low pressure side), a WISP 712 autosampler, a Lambda-Max 481 variable wavelength UV/VIS monitor, and a 745 data module (all fromWaters). The CoA derivatives were detected at 260 nm after separation on a Merck LiChrospher 100 RP18-column (5 μm, 125 × 4 mm, precolumn 4 × 4 mm, LiChroCART system).

Herbicides

The herbicides used were sethoxydim (2-(1-(ethoxyimino)butyl)-5-(2-(ethylthio)propyl)-3-hydroxy-2-cyclohexene-1-one), cycloxydim (2-(1-(ethoxyimino)-butyl)-3-hydroxy-5-((2H)-tetrahydrothiopyran-3-yl)-2-cyclohexene-1-one), and diclofop (2-(4-(2′,4′-dichlorophenoxy)-phenoxy)propanoate).

RP18 HPLC of CoA Derivatives

The different CoA derivatives were separated by the following binary gradient (buffer A + buffer B = 100%): 7 min isocratic 0% buffer B, 1 mL/min; 3 min linear gradient to 70% buffer B, 2 mL/min; 3 min isocratic 70% buffer B, 2 mL/min; 3 min linear gradient to 0% buffer B, 2 mL/min; 4 min linear gradient to 0% buffer B, 1 mL/min (buffer A: 92% 100 m*M* KH_2PO_4, pH 5.0, 8% methanol; buffer B: 50% 100 m*M* KH_2PO_4, pH 5.0, 50% methanol). After at most 50 runs, the column was regenerated with either 100% buffer B or with pure methanol. The absorbance was recorded at 260 nm.

Determination of I_{50} Values

In order to determine the I_{50} values (i.e., the concentration of the inhibitor which inhibits 50% of the enzyme activity), ACCase preparations with comparable specific activity were used. The inhibitor solutions were diluted in steps of 1:10, and the enzyme activity was measured by quantitative analysis of the malonyl-CoA peak. The degree of ACCase-inhibition was calculated in relation to the maximum activity of the enzyme and its maximum inhibition by 1 m*M* sethoxydim. The I_{50} values were calculated by linear interpolation between the two relevant inhibitor concentrations.

RESULTS

HPLC Analysis of CoA Derivatives

A typical HPLC spectrum of malonyl- and acetyl-CoA standards recorded at 260 nm is shown in Figure 1. With the method described above, it presents no problem to detect 1 μg of a CoA derivative. Figure 2 shows the result of an enzyme assay started with aceytl-CoA. Malonyl-CoA and acetyl-CoA can be analyzed in one HPLC run. This, together with the regeneration of the column, takes 20 min. With an autosampler 72 samples can be handled per day. One HPLC column can be used for at least 1000 injections if the buffers and probes are treated properly. The precolumn is changed after 300 injections.

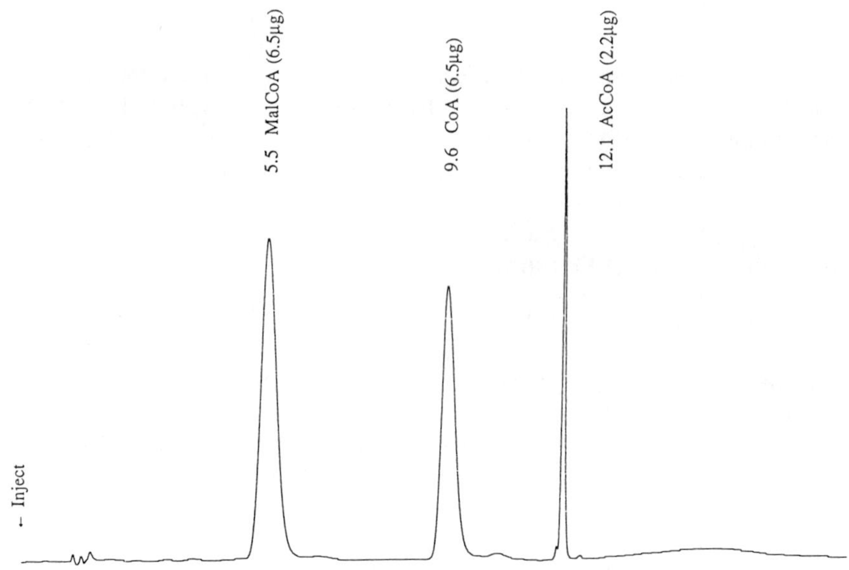

Figure 1. HPLC chromatogram illustrating the separation of acetyl-CoA (AcCoA), malonyl-CoA (MalCoA), and CoA standards (conditions as described in the text).

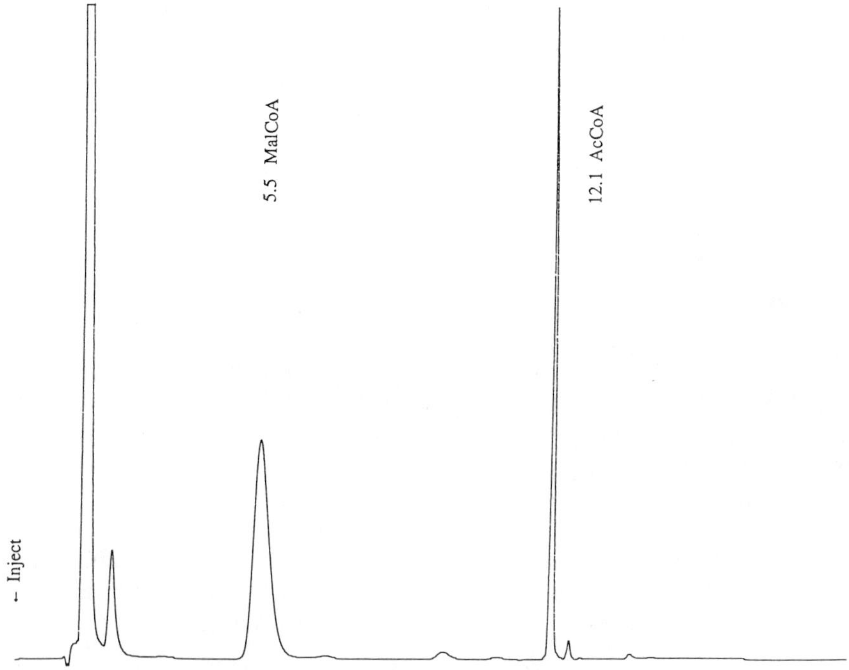

Figure 2. HPLC chromatogram illustrating acetyl-CoA carboxylase activity in corn preparations (conditions as described in the text).

Table 1. Comparison of I_{50} Values of Different Herbicides Evaluated either by the $^{14}CO_2$ Incorporation Method or by the HPLC Method.

Herbicide	I_{50} ($^{14}CO_2$) (μM)	I_{50} (HPLC) (μM)
Cycloxydim	0.5	0.8
Sethoxydim	2.0	3.0
Diclofop	0.1	0.5

I_{50} Values of Herbicides

Table 1 shows the I_{50} values of sethoxydim, cycloxydim, and diclofop. These results come very close to inhibition constants evaluated by the $^{14}CO_2$ incorporation method. In general, inhibitors do not interfere with the malonyl-CoA or acetyl-CoA peak. No problems were experienced with plant ingredients in the HPLC analysis so long as the enzyme was prepared as described above.

COMMENTS

Using the method described here, it is possible to analyze acetyl-CoA, the substrate, and malonyl-CoA, the product of the enzyme reaction. Moreover, the activity of esterases (possible contaminants in the enzyme preparations) can be detected by a decrease of the acetyl-CoA peak and a corresponding increase of the CoA peak. This is important because these esterases can influence the enzyme kinetics by degrading the substrate to a concentration which is insufficient for optimal enzyme activity. It should be mentioned again that there is no need to work with radioactive material, and it is not necessary to trap the $^{14}CO_2$, which is evaporated after acidifying the enzyme solution. The only problems with the HPLC method are that the number of possible tests per day is smaller than with the radioactive method and that some inhibitors might interfere with the malonyl-CoA peak.

REFERENCES

Burton, J. D., Gronwald, J. W., Somers, D. A., Connelly, J. A., Gengenbach, B. G. and Wyse, D. L., Inhibition of plant acetyl-coenzyme A carboxylase by the herbicides sethoxydim and haloxyfop. *Biochem. Biophys. Res. Commun.*, 148, 1039, 1987.

Burton, J. D., Gronwald, J. W., Somers, D. A., Gengenbach, B. G. and Wyse, D. L., Inhibition of corn acetyl-coenzyme A carboxylase by cyclohexanedione and aryloxyphenoxypropionate herbicides. *Pestic. Biochem. Physiol.*, 34, 76, 1989.

Corkey, B.E., Brandt, M., Williams, R. J. and Williamson, J. R., Assay of short-chain acyl coenzyme A intermediates in tissue extracts by HPLC. *Anal. Biochem.*, 118, 30, 1981.

Egin-Bühler, B., Loyal, R. and Ebel, J., Comparison of acetyl-CoA carboxylases from parsley cell cultures and wheat germ. *Arch. Biochem. Biophys.* 203, 90, 1980.

Focke, M. and Lichtenthaler, H. K., Inhibition of the acetyl-CoA carboxylase of barley chloroplasts by cycloxydim and sethoxydim. *Z. Naturforsch.*, 42c, 1361, 1987.

CHAPTER 23

A Novel Procedure for Measurement of Acetyl-CoA Carboxylase in Comparison with the ^{14}C Fixation Method

Joy L. Howard and Stuart M. Ridley

INTRODUCTION

Acetyl-CoA carboxylase (acetyl-CoA carbon dioxide ligase (ADP-forming); EC 6.4.1.2) is the first step committed to the pathway of fatty acid biosynthesis, converting acetyl-CoA to malonyl-CoA. The enzyme from plants is the site of action of two separate chemical groups of grass-selective herbicides.

1. The aryloxyphenoxypropionic acids are inhibitory to the monocot leaf enzyme, with considerably less activity against the dicot enzyme; only the R-enantiomer inhibits the enzyme, and kinetic analyses have shown noncompetitive inhibition with respect to all three substrates (acetyl-CoA, MgATP, and bicarbonate) by one of the herbicides (diclofop) in this series (Rendina et al., 1988). Diclofop is (RS)-2[4-(2,4-dichlorophenoxy)phenoxy]propionic acid.
2. The cyclohexanedione herbicides also inhibit acetyl-CoA carboxylase with a greater inhibition of the enzyme from monocots than dicots. Thus, the primary basis for selectivity by both chemical types appears to be the extent to which acetyl-CoA carboxylase is inhibited in the two groups of plants.

0-87371-539-X/93/$0.00 + $.50

The enzymic reaction is a two-step process. The first step involves the ATP-dependent carboxylation of biotin in a biotin carboxyl carrier protein (BCCP) by biotin carboxylase, which results in the liberation of ADP and phosphate. In the second step the carboxyl group is transferred to acetyl-CoA by transcarboxylase activity to produce malonyl-CoA:

$$\text{ATP} + \text{CO}_2 + \text{BCCP} \rightarrow \text{CO}_2\text{-BCCP} + \text{ADP} + \text{P}_i$$

$$\text{CO}_2\text{-BCCP} + \text{acetyl-CoA} \rightarrow \text{malonyl-CoA} + \text{BCCP}$$

The enzyme from plants can be dissociated into high-molecular weight subunits of approximately 220,000 Da, but with three catalytic domains present on each polypeptide subunit. As a consequence of this, and also the two-site kinetic mechanism proposed by Finlayson and Dennis (1983), it could be anticipated that the two reaction steps might remain tightly coupled and so allow the use of the production of phosphate as an easily manageable assay procedure, particularly for rapid kinetic analyses.

The highly sensitive phosphate assay has been compared with the more widely employed ^{14}C-bicarbonate assay to determine acetyl-CoA k_m values and inhibition kinetics by the herbicide fluazifop.

PROCEDURE

Equipment

For the ^{14}C-bicarbonate assay, a liquid scintillation counter is required for final estimation of results. The inorganic phosphate assay requires a spectrophotometer which measures absorbance at 630 nm. For both assays, a thermostatically controlled heating block is required.

Isolation of Acetyl-CoA Carboxylase from Maize

Acetyl-CoA carboxylase was extracted from freshly harvested 10-day-old maize (*Zea mays* var. Pioneer 3183) leaves which were homogenized in a Waring blender in ice-cold Buffer A, pH 8.0, containing 0.1 *M* Tris, 20 m*M* mercaptoethanol, 1 m*M* EDTA, 0.5% Triton X-100, 1 m*M* benzamidine, 10 m*M* magnesium chloride, and 20% glycerol. The extract was centrifuged at 27,500 × *g* for 30 min, and the resulting supernatant was filtered through 50-μm filter cloth followed by precipitation with ammonium sulfate between 35 and 45%. The final pellet was resuspended in Buffer B, pH 8.0, containing 10 m*M* Tris, 20 m*M* and mercaptoethanol, 1 m*M* EDTA, 1 m*M* benzamidine, 10 m*M* magnesium chloride, and 20% glycerol, and passed down a gel filtration column containing Sephacryl S300HR pre-equilibrated with Buffer B but containing only 10% glycerol. Active fractions were pooled and loaded onto a preparative Mono Q ion exchange column (attached to an FPLC system), previously equilibrated with buffer containing 10 m*M* Tris, 1 m*M* dithiothreitol, 1 m*M* EDTA, 1 m*M* benzamidine, 10 m*M* magnesium chloride, and 10% glycerol, pH 8.0. Acetyl-CoA carboxylase was eluted with a KCl gradient between 0.2 and 0.5 *M*. Protein concentrations of enzyme preparations were measured using the method of Bradford (1976).

^{14}C-Bicarbonate Assay

During the isolation procedure the enzyme was assayed by the method of Nikolau et al. (1981). A final assay volume of 200 μL contained 0.1 *M* Tricine-KOH (pH 8.0), 0.5 m*M* ATP, 2.5 m*M* magnesium chloride, 50 m*M* potassium chloride, 30 m*M* $NaH^{14}CO_3$ (0.27 μCi), 1 m*M* dithiothreitol, 0.3 m*M* acetyl-CoA (omitted from controls), and up to 100 μL enzyme preparation. Samples were preincubated at 30°C for 3 min prior to addition of acetyl-CoA to start the reaction, followed by vigorous mixing and incubation at 30°C for 10 min. The reaction was stopped by the addition of 50 μL 6 *M* HCl, followed by further vigorous mixing. For the ^{14}C assay, a 75-μL aliquot was removed and dried on a 3.2-cm-diameter Whatman GF/A filter. The radioactivity was counted in Optiphase scintillation fluid in a Packard 460 scintillation counter. In view of the difficulties of calibrating disintegrations per minute with nanomoles malonyl-CoA, all results of the radioactive assay are given as dpm incorporated into malonyl-CoA.

Inorganic Phosphate Assay

In establishing an alternative assay procedure, the remaining incubation mixture was employed for phosphate determination using a modified method based on that of Hess and Derr (1975) and Howard and Ridley (1990). A Malachite Green solution was prepared by mixing three parts 0.045% Malachite Green hydrochloride in distilled water with one part 4.2% ammonium molybdate in 4 *M* HCl. The solution was left to stand for 30 min, then filtered through an "Acrodisc" 0.45-μm PTFE (polytetrafluoroethylene) filter. Triton X-100 (1.5%) in distilled water was added to a final concentration of 0.0375%, and 750 μL of this solution were added to the remaining incubation mixture. After allowing this mixture to stand for 3 min, 150 μL 34% citric acid were added, and samples were read at 630 nm after standing for a further 60 min. Experimental samples were compared with standards containing 1 to 10 nmol inorganic phosphate.

Herbicide Inhibition Experiments

Inhibition experiments were carried out using the active R-enantiomer of fluazifop. This is the active acid component of the grass-specific herbicide fluazifop-butyl that has been shown to inhibit acetyl-CoA carboxylase in barley, but not in peas. Results of inhibition experiments are given in terms

Table 1. Partial Purification of Maize Acetyl-CoA Carboxylase

Enzyme Fraction	Specific Activity (dpm/mg)	% Recovery	Overall Purification
1. Crude extract	1,125	—	—
2. $(NH_4)_2SO_4$ precipitate	1,913	26	1.7
3. Sephacryl S300 HR	36,610	22	33
4. Mono Q peak 2	203,309	1.92	181

of IC_{50}. This is defined as the concentration of inhibitor required to give 50% inhibition. IC_{50} values and K_m values were calculated using a "FIT" program (Dr. D. K. Lawrence, © ICI Agrochemicals, Jealott's Hill, Bracknell, Berkshire, England).

RESULTS

The activity of acetyl-CoA carboxylase during purification from maize shoots was followed using the ^{14}C assay described. This gave a measure of the overall purification factor (Table 1). After elution from the Mono Q column, two peaks of enzyme activity were seen with the ^{14}C assay. It was established that acetyl-CoA carboxylase activity could also be determined by measuring the liberation of inorganic phosphate in the presence of acetyl-CoA, and two peaks of activity were confirmed by this procedure as well. With both assays, activity was localized in the same fractions (Figure 1), and measurement with the phosphate assay gave a specific activity in fraction 15 of 173 nmol/min/mg. The first peak of acetyl-CoA carboxylase activity (peak 1) coincided with a major protein peak which also contained nonspecific phosphatases (seen as an increase in phosphate production in the absence of acetyl-CoA).

Analysis of the Michaelis-Menten kinetics for acetyl-CoA carboxylase activity in peak 2 with respect to acetyl-CoA gave K_m values of 0.236 mM (± 0.060, n = 4) for the phosphate assay, and 0.279 mM (± 0.065, n = 3) for the ^{14}C assay. Both assays showed inhibition of peak 2 acetyl-CoA carboxylase by fluazifop, and dose-response curves using both assays gave similar IC_{50} values. (Peak 2 IC_{50} by the phosphate assay was 4.99 μM; by the ^{14}C assay, it was 2.45 μM.)

Kinetics for inhibition by fluazifop were also investigated with both assays using acetyl-CoA carboxylase activity in peak 2 and acetyl-CoA as the variable substrate (Figures 2 and 3); in both experiments inhibition was noncompetitive.

COMMENTS

With the recent increased interest in acetyl-CoA carboxylase from plant sources as the site of action of new grass-selective herbicides, an assay procedure more readily accessible than the ^{14}C assay has been sought to make

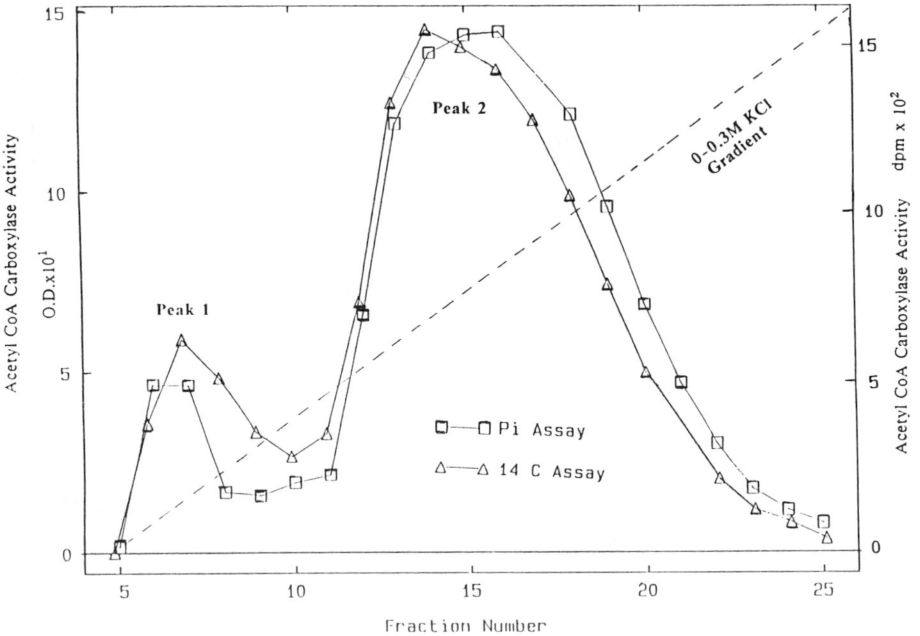

Figure 1. Purification of maize acetyl-CoA carboxylase using Mono Q, measured using ^{14}C and phosphate assays.

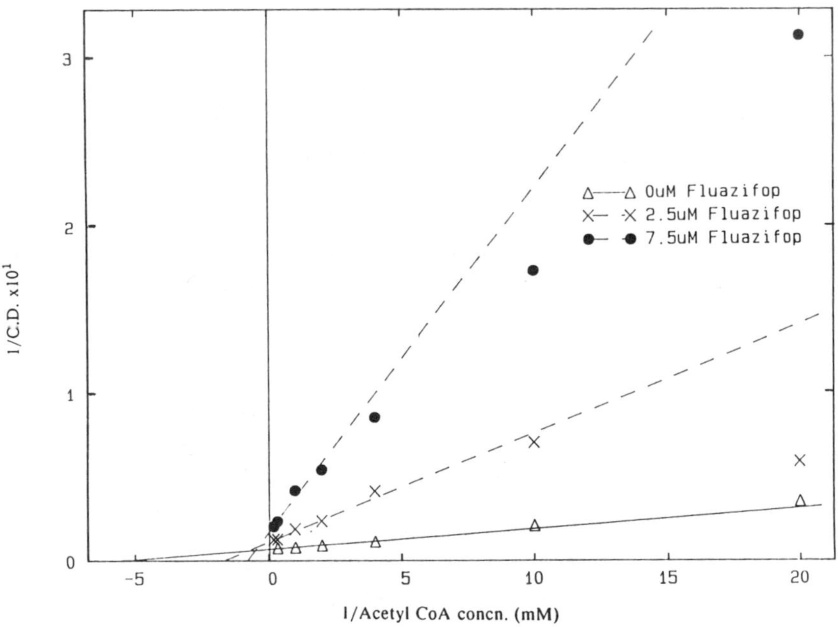

Figure 2. Inhibition kinetics for fluazifop acid, with acetyl-CoA as the variable substrate, measured using the inorganic phosphate (Pi) assay.

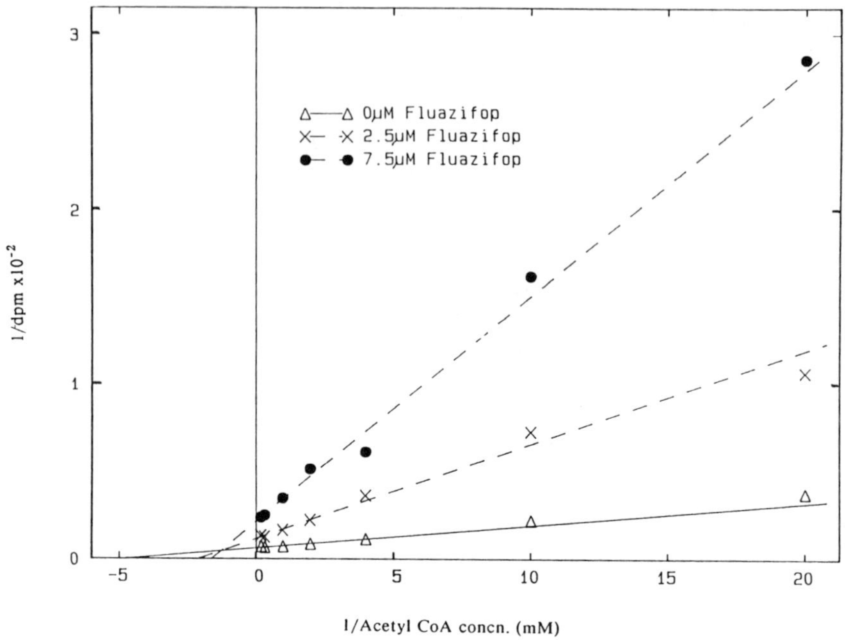

Figure 3. Inhibition kinetics for fluazifop acid, with acetyl-CoA as the variable substrate, measured using the ^{14}C assay.

kinetic analyses easier. Acetyl-CoA carboxylase activity, purified from maize 180-fold, has been assayed by ^{14}C fixation and by phosphate release. Figure 1 shows a close correlation between the two procedures for the activity of acetyl-CoA carboxylase eluting from an ion exchange column. Both methods registered major (peak 2) and minor (peak 1) peaks of activity, with the purer product being in peak 2.

Since phosphate is liberated as a product of the first half of the reaction sequence, the two reactions must remain tightly coupled during purification, and both reactions must be present in the two peaks of activity. Measurement of K_m values for acetyl-CoA by the two assay procedures using the activity in peak 2 gives closely similar results.

Inhibition kinetics by the two assays both indicate noncompetitive inhibition of acetyl-CoA carboxylase in peak 2, and the IC_{50} values at 3 to 5 μM show no significant difference. It is therefore clear that the total reactions that constitute activity of acetyl-CoA carboxylase can be recorded reliably using the liberation of phosphate in the first half of the reaction sequence. It may be anticipated that the specific activity of the enzyme measured by phosphate release is more accurate than that measured by ^{14}C fixation because of the nonspecific absorption of CO_2 in the reaction due to the partial pressure of carbon dioxide in the atmosphere. The use of the phosphate assay provides advantages over the ^{14}C assay in the speed of analysis and in accuracy, although the former suffers in crude plant extracts, where high levels of nonspecific phosphatase activity may be present.

REFERENCES

Rendina, A. R., Felts, J. M., Beaudoin, J. D., Craig-Kennard, A. C., Look, L. L., Paraskos, S. L. and Hagenah, J. A., Kinetic characterization, stereoselectivity, and species selectivity of the inhibition of plant acetyl-CoA carboxylase by the aryloxyphenoxypropionic acid grass herbicides. *Arch. Biochem. Biophys.*, 265, 219, 1988.

Finlayson, S. A. and Dennis, D. T., Acetyl-coenzyme-A a carboxylase from the developing endosperm of *Ricinus communis*. *Arch. Biochem. Biophys.*, 225, 586, 1983.

Bradford, M. M., A rapid and sensitive method for the quantitation of microgram quantities of protein utilizing the principle of protein-dye binding. *Anal. Biochem.*, 72, 248, 1976.

Nikolau, B. J., Hawke, J. C. and Slack, C. R., Acetyl-coenzyme A carboxylase in maize leaves. *Arch. Biochem. Biophys.*, 211, 605, 1981.

Hess, H. H. and Derr, J. E., Assay of inorganic and organic phosphorus in the 0.1–5 nanomole range. *Anal. Biochem.*, 63, 607, 1975.

Howard, J. L. and Ridley, S. M., Acetyl-CoA carboxylase: a rapid novel assay procedure used in conjunction with the preparation of enzyme from maize leaves. *FEBS Lett.*, 261, 261, 1990.

CHAPTER 24

Changes in Fatty Acid Profile Induced by Herbicides

Michel Couderchet and Peter Böger

INTRODUCTION

Two sites of herbicidal inhibition have been described in plant fatty acid biosynthesis. The first and best known of both sites is at the very early stage of the biosynthesis, where cyclohexanediones inhibit the acetyl-CoA carboxylase (see Chapters 21 to 23). At the other end of the biosynthesis the desaturation of the C16:0* and C18:1 can be influenced by herbicides. The consequence of this is a possible change in membrane permeability and fluidity and thereby a misfunction of the plant leading to death. Pyridazinone derivatives like norflurazone, metflurazone, or BAS 13338 (SAN 9785) were shown to induce an increase in the C18:2/C18:3 ratio *in vivo* (for review see Hoppe, 1989). Recently, BAS 13338 was found to inhibit the C18:2 desaturase activity in an *in vitro* enzymatic test (Norman et al., 1991). Desaturation of fatty acids can also be inhibited by chloroacetamides like alachlor and metazachlor (5 μM), which dramatically reduce the proportion of polyunsaturated fatty acids in *Scenedesmus acutus* cells (Weisshaar et al., 1988).

* For the fatty acid Cx:y, x represents the number of C atoms and y the number of double bonds.

0-87371-539-X/93/$0.00 + $.50

By gas chromatography, a fatty acid profile is relatively easy to determine and may provide some valuable information on the target of a phytotoxic compound. In this chapter, a method to rapidly assess the influence of herbicides on the fatty acid profile of *S. acutus* and *Glycine max* (soybean) cells is described.

PROCEDURES

Plant Material

The green alga *Scenedesmus acutus* no. 276-3a of the Algae Collection, University of Göttingen, Göttingen, Germany was grown in 200 mL of a liquid medium containing (per liter) KNO_3 (1 g), $Na_2HPO_4 \cdot 2H_2O$ (0.26 g), KH_2PO_4 (0.74 g), $MgSO_4 \cdot 7H_2O$ (0.05 g), $CaCl_2$ (7.56 mg), $Fe(NO_3)_3 \cdot 9H_2O$ (14.5 mg), Tritiplex III (13.3 mg), $Al_2(SO_4)_3 \cdot 18H_2O$ (6.6 μg), $MnCl_2 \cdot 4H_2O$ (99 μg), $CoSO_4 \cdot 7H_2O$ (2.8 μg), $ZnSO_4 \cdot 7H_2O$ (6.3 μg), H_2BO_3 (30.6 μg), $(NH_4)_6Mo_7O_{24} \cdot 4H_2O$ (1.8 μg), NH_4VO_3 (2.9 μg), $NiSO_4 \cdot 6H_2O$ (26.3 μg), KI (0.25 mg), KBr (0.24 mg), LiCl (8.4 μg), and $CuSO_4 \cdot 5H_2O$ (2.5 μg). The cultures were inoculated with a packed cell volume (pcv) of 0.3 μL cell per milliliter suspension (see Chapter 7 for determination of pcv). Light (80 μEinstein (μE) $\times$ m^{-2} $\times$ sec^{-1}) was provided by fluorescent lamps, temperature was 21°C, and a continuous flow of air supplemented with 2% (v/v) CO_2 was bubbled through the culture.

Soybean cells were grown in mixotrophic conditions in Erlenmeyer flasks as described in detail in Chapter 8. Cultures were inoculated with 15 mL of a 5-day-old cell suspension in 35 mL of fresh medium.

Equipment

Basic equipment for growth of plant material included a gas chromatograph (equipped with a flame ionization detector and a column adequate to separate fatty acid methyl esters) and thermostated water bath (65°C).

Herbicides

BAS 13338 (4-chloro-5-(dimethylamino)-2-phenyl-3(2*H*)-pyridazinone) and metazachlor (2-chloro-*N*-(2,6-dimethylphenyl)-*N*-1*H*-pyrazol-1-ylmethyl)-acetamide, 99.6% pure) were generous gifts from BASF AG (Limburgerhof, Germany). Alachlor (2-chloro-*N*-(2,6-dimethylphenyl)-*N*-(methoxymethyl)-acetamide, 99.3% pure) was purchased from Dr. Ehrenstorfer (Augsburg, Germany). Stock solutions were made in ethanol so that the concentration of ethanol in the treatments was 0.1 or 0.2%. Herbicides were added to the cultures at inoculation time.

Extraction of Fatty Acids

Algal (10 mL) or cell (1 mL) suspension was centrifuged (5 min, approximately 6000 × *g*). The pellet was washed and resuspended in 1 mL water. Methanolic KOH (10% w/v, 5 mL) was added. The tubes were capped and incubated at 65°C for 90 min to allow for total saponification of the lipids. The samples were cooled and the carotene fraction removed by washing twice with 3 mL *n*-hexane. After acidifying (1 mL of 37% HCl), the fatty acids were extracted twice with 3 mL *n*-hexane.

Fatty Acid Methylation

In order to be separated by gas chromatography, the fatty acids have to be methylated. The fatty acid fraction is dried under nitrogen. The dried fatty acids are taken up in 500 μL boron trifluoride (14% in MeOH) and incubated 5 min at 100°C in a tightly capped vessel. After cooling, 250 μL water is added and the fatty acid methyl esters are finally extracted with 500 μL *n*-hexane.

Gas Chromatography of the Fatty Acid Methyl Esters

Fatty acid methyl esters were separated using a gas chromatograph (Perkin-Elmer P22) equipped with a flame ionization detector (FID). The carrier gas (nitrogen) supply was 30 mL/min. Injector and detector temperature was 250°C, and that of the oven was 165°C. Separation was performed using a fused silica capillary column (SP-2330, 30 m × 0.25 mm, Supelco, Bad Homburg, Germany) or a used packed stainless steel column (15% DEGS, diethylene glycol succinate, on Chromosorb W AW, Manville, Corp., with 80 to 100 mesh, 2 m × 1/8 in., Supelco, Bad Homburg, Germany). The instrument was equipped with a Hewlett Packard (Model 2220) recording integrator. A 200-μL volume of the fatty acid methyl esters was concentrated to 20 μL under a stream of nitrogen, and 2 μL (steel column) or 0.1 μL (capillary column) were injected into the gas chromatograph.

Determination of Fatty Acids

Identification was based on commercially available reference methyl esters for the most common fatty acids. Unusual fatty acids (C16:2 to C16:4 and C18:4) were identified by comparing with the literature, after separation by silver nitrate column chromatography (Christie, 1989) and/or were confirmed by mass spectrometry.

RESULTS

Qualitative observations can be made directly on the chromatogram or after peak integration. The most common fatty acid in soybean mixotrophic cells

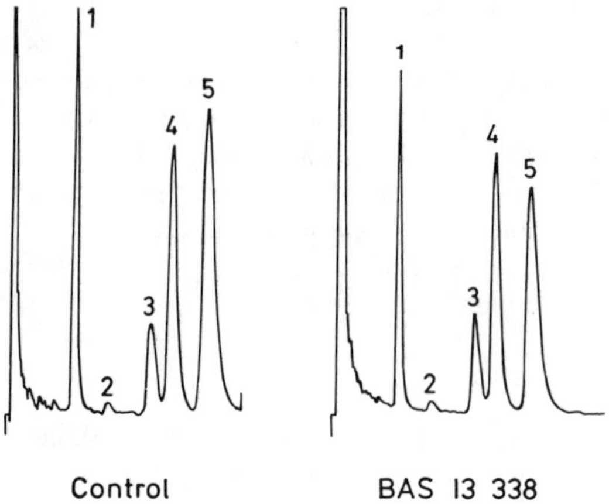

Figure 1. Fatty acid profile of soybean cells after a 72-hr incubation with or without BAS 13338 (200 μ*M*). The fatty acid extract (2 μL) was injected into a gas chromatograph equipped with a packed stainless steel column. Peaks 1 to 5 correspond to C16:0, C16:1, C18:0–1 (C18:0 and C18:1 coeluted), C18:2, and C18:3, respectively.

was C18:3 (peak 5 of control, Figure 1). It represented more than 50% of the total area below the chromatogram. After incubation with BAS 13338 the amount of C18:3 was reduced, the sizes of peaks 4 and 5 (which represent C18:2 and C18:3, respectively) were inverted, and C18:2 became the most important (Figure 1). This result was confirmed after integration of the results, giving 31.3% C18:3 and almost 40% C18:2 (Table 1). The influence of alachlor and metazachlor on the fatty acid profile was much smaller; C18:3 only decreased slightly while C18:2 accumulted. This could be expected since soybean is resistant to chloroacetamides. Using the packed column C18:0 and C18:1 coeluted, while separation with a capillary column gave equal amounts of both fatty acids in untreated cells (see Figure 3).

In *S. acutus* the major fatty acid was also C18:3; in addition, many other fatty acids were represented (Figure 2, Table 2). The fatty acid profile of this algae was strongly influenced by alachlor (Figure 2). After 24 hr, peaks 8 to 10 (C18:2 to C18:4) were smaller with 5 μ*M* alachlor and the C18:1 peak (peak 6) was much larger. On the chromatogram of the control cells the C16:4

Table 1. Influence of Herbicides (200 μ*M*) on the Fatty Acid Distribution (in Percent of Total Fatty Acids) of Soybean Cells Incubated for 144 hr

Fatty Acid	Control	Alachlor	Metazachlor	BAS 13338
C16:0	22.1	19.6	19.3	19.6
C16:1	0.9	0.6	1.8	0.4
C18:0–1	4.9	6.5	6.0	8.8
C18:2	19.8	28.0	26.2	39.8
C18:3	52.4	45.3	46.8	31.3

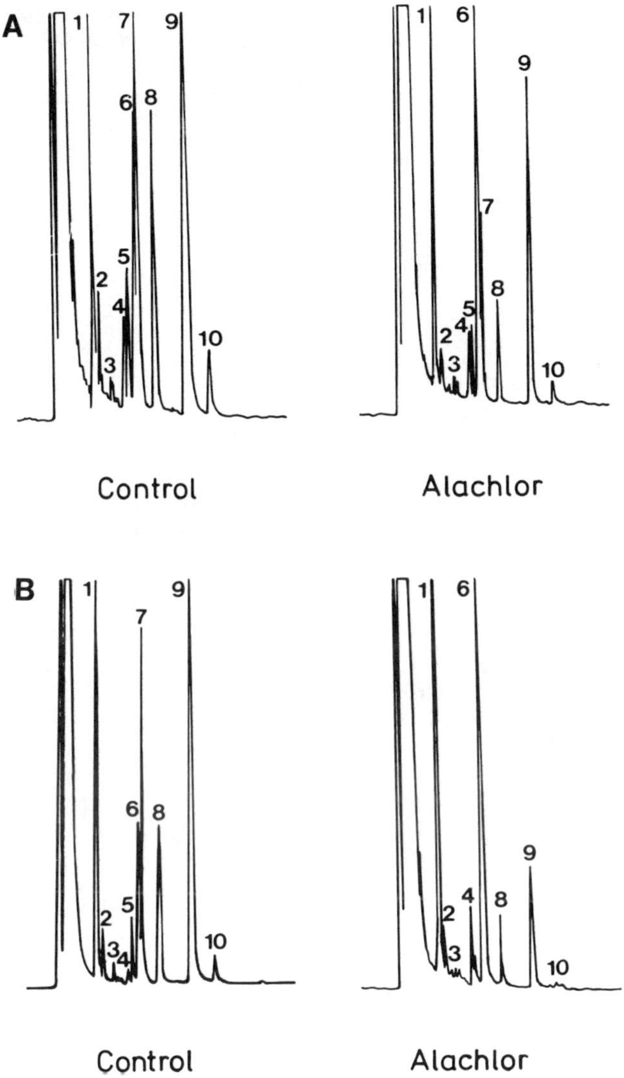

Figure 2. Fatty acid profile of *Scenedesmus acutus* cells after incubation with or without alachlor (5 μM). The gas chromatograph was equipped with a fused silica capillary column. The peak numbers correspond to those in Table 2. (A) 24-hr incubation; (B) 40-hr incubation.

peak (peak 7) is higher than that of C18:1 (peak 6); this situation is inverted with alachlor after 24 hr (Figure 2A). After 40 hr the C16:4 peak has disappeared (Figure 2B). The effect of metazachlor was similar to that of alachlor (Table 2). In contrast, BAS 13338 had almost no effect on the fatty acid profile of this algae (Table 2).

Table 2. Influence of Herbicides (5 μ*M*) on the Total Fatty Acid Distribution (in Percent of Total Fatty Acids) of *Scenedesmus acutus* Cells Incubated for 40 hr.[a]

Fatty Acid	Peak Number	Control	Alachlor	Metazachlor	BAS 13338
C16:0	1	21.5	32.8	28.4	20.4
C16:1	2	2.6	2.8	2.1	3.2
C16:2	3	0.9	0.8	0.7	1.2
C16:3	5	2.5	0.8	1.0	2.8
C16:4	7	12.6	tr.	tr.	14.1
C18:0	4	0.2	1.4	1.0	0.5
C18:1	6	0.2	34.9	42.6	1.5
C18:2	8	12.1	6.8	5.8	13.3
C18:3	9	44.7	18.5	17.4	40.5
C18:4	10	2.8	1.2	1.0	2.5

[a] The peak numbers correspond to those on the chromatogram (see Figure 2); tr. = trace amounts.

The results indicate relative amounts of the various fatty acids, but it is also possible to obtain absolute concentrations for the individual fatty acids. Adding a given quantity of a fatty acid absent in the sample (e.g, C17:0) before injecting the fatty acid mixture in the gas chromatograph makes it possible to calibrate the area integrated. A soybean fatty acid sample was injected into the gas chromatograph after 10 μL of a heptadecanoic acid methyl ester (C17:0) solution (41 $\mu g \times \mu L^{-1}$) were added to 10 μL of the soybean fatty acid methyl ester mixture. The C17:0 peak represented 42% of the total peak area, which in turn corresponded to 10.25 μg of the fatty acid. The amount of the other fatty acids could then be calculated since it is proportional to the area integrated (Figure 3).

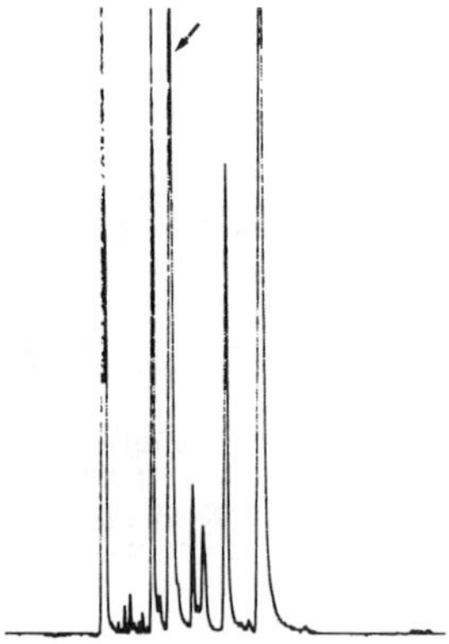

Figure 3. Fatty acid profile of soybean cells after C17:0 (arrow) was added to the fatty acid mixture. The gas chromatograph was equipped with a fused silica capillary column. The relative quantities of C16:0, C17:0, C18:0, C18:1, C18:2 and C18:3 were 11.6, 42.3, 1.7, 2.0, 8.5, and 33.9%, respectively, which corresponded to the following injected absolute quantities: 2.81, 10.25, 0.41, 0.48, 2.06, and 8.21 μg, respectively.

COMMENTS

The relatively simple determination of the fatty acid profile gives an indication of the mode of action of phytotoxic compounds. The decrease of C18:3 in soybean and corresponding increase in C18:2 indicate that the desaturation step between C18:2 and C18:3 is inhibited. In cells treated with chloroacetamides, the amount of C18:3 decreased: however, in that case C18:2 and C18:4 also decreased and C18:1 accumulated. From the fatty acid profile it can be concluded that both BAS 13338 and chloroacetamides inhibit fatty acid desaturation, but their target site is different.

The method described for cell suspension can be adapted easily to plant tissue in the following manner. A small amount of tissue is ground in a mortar and washed before saponification is allowed to proceed at 65°C for 90 min. The rest of procedure is identical to that described above for cells.

The column should be adapted to the fatty acids of the plant species studied. For example, the used packed column would not be adequate for the separation of *S. acutus* fatty acids. Indeed, in that case, C16:4 would coelute with C18:0 and it would lead to erroneous conclusions on the influence of chloroacetamide herbicides. This column was sufficient for the separation of soybean fatty acids.

REFERENCES

Christie, W. W., Silver ion chromatography using solid-phase extraction columns packed with a bonded-sulfonic acid phase. *J. Lipid Res.*, 30, 1471, 1989.

Hoppe, H. H., Fatty acid biosynthesis. A target site of herbicide action, in *Target Sites of Herbicide Action,* Böger, P. and Sandmann, G., Eds., CRC Press, Boca Raton, FL, 1989, p. 65.

Norman, H. A., Pillai, P., and St. John, J. B., In vitro desaturation of monogalactosyldiacylglycerol and phosphatidylcholine molecular species by chloroplast homogenates. *Phytochemistry,* 30, 2217, 1991.

Weisshaar, H., Retzlaff, G. and Böger, P., Chloroacetamide inhibition of fatty acid synthesis. *Pestic. Biochem. Physiol.*, 32, 212, 1988.

Inhibition of Cell Division

CHAPTER 25

Components of Plant Growth and Its Measurement

F. Dan Hess and Thomas L. Rost

INTRODUCTION

Many preemergence herbicides exhibit growth inhibition as their primary mode of action. If growth inhibition prevents a germinating weed from emerging above the soil surface, it is effectively controlled even if no other physiological event (e.g., membrane disruption) occurs. Carbohydrate and lipid reserves present in the endosperm are finite and cannot sustain the seedling indefinitely. Preemergence herbicide classes that have growth inhibition as a primary mode of action are dinitroanilines, phosphorothioamidates, phenylcarbamates, thiocarbamates, choroacetamides, sulfonylureas, and imidazolinones. Herbicide interference at a variety of biochemical target sites will result in growth inhibition (see Hess, 1982 and 1989, for review); thus, observing growth inhibition reveals little about the mechanism of action.

Plant growth can be defined as a combination of cell division and cell enlargement that leads to an irreversible increase in size. Cell division without subsequent cell enlargement does not result in an increase in size, and cell enlargement within an individual cell is finite. Herbicide inhibition and/or disruption of either process will result in growth inhibition. Subsequent chapters deal with measuring inhibition and disruption within cell division (the cell cycle) and cell enlargement. This chapter defines a simple test that can

0-87371-539-X/93/$0.00 + $.50

be used to determine if growth inhibition is a primary herbicide mode of action.

PROCEDURES

The most important parameters in analyzing the effect of an herbicide on plant growth are dose (concentration) and duration of treatment. Growth inhibition studies must include a wide range of herbicide concentrations and treatment times in order to determine if the observed affect on growth is related to a primary or secondary mode of action.

The concentration range used should be centered around the use rate needed to control weeds in the field. Although it is not possible to precisely convert field use rates to laboratory concentrations, approximations are possible. Using a furrow slice approximation, 1 ha of soil, 7.5 cm deep (an assumed herbicide zone), weighs approximately 1×10^6 kg. Although variable among soil types, an average water content in agricultural soils is 30 to 40% of the soil dry weight; thus, the water in 1 ha of soil, 7.5 cm deep, would be 3×10^5 to 4×10^5 kg. If a herbicide application of 1 kg/ha becomes dissolved in the water, the concentration would be approximately 2 to 3 ppm. Herbicide binding to soil will reduce this concentration to 1 ppm or less. Therefore, ranges of concentrations to use in a soil model system should be between 0.01 and 100 ppm for herbicides applied at 0.1 to 1.0 kg/ha. A broad concentration range should be used in initial experiments, followed by a narrow range centered around the concentration where effects of the herbicide were first observed. To compare among intrinsic activity of different herbicides it is useful to change the units of measure to molar.

For herbicides that inhibit growth as a primary mechanism, the inhibition is observed within hours. Therefore, the length of treatment durations in initial experiments need not be longer than 48 to 96 hr. Initial experiments can evaluate intervals of 6 to 12 hr between measurements to determine the approximate time growth inhibition begins. In order to determine a more precise time, follow-up experiments may require measurement intervals between 3 and 6 hr.

Preemergence herbicides that inhibit growth exert their effect on germinating seedlings prior to emergence from the soil; therefore, the most appropriate method for measuring growth inhibition is a model soil system. The most suitable substrate to replace soil in the model system is washed silica sand. This substrate is considered inert in that essentially no water or herbicide adsorbs or absorbs. First, square plastic petri dish bottoms ($9 \times 9 \times 1.34$ cm) are completely filled with dry (but previously washed) silica sand. Then, 35 mL of one-quarter-strength nutrient solution (e.g., Hoagland and Arnon, 1950) containing the herbicide are added to each petri plate. To simulate a soil moisture of ''field capacity'', excess nutrient solution remaining on the bottom of the petri dish is removed with a Pasteur pipette.

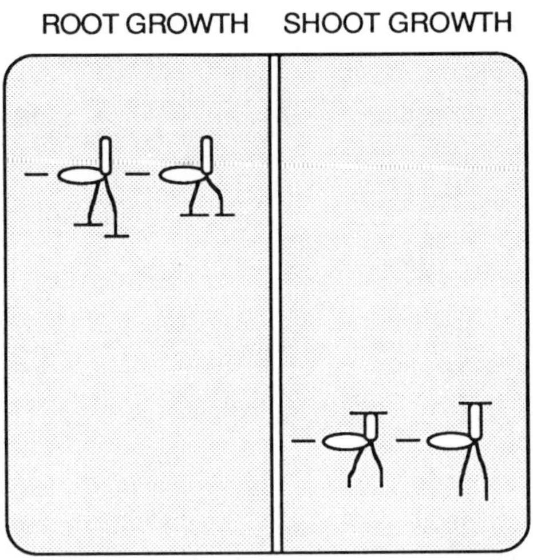

Figure 1. Diagram of seedling placement on the sand surface for root and shoot growth experiments. At the start of the incubation period the location of the seed, root tip for root growth, and shoot tip for shoot growth are marked on the petri dish cover.

Small-seeded broadleaf and grass species work best in the culture system described below; however, species with seeds as large as peas have been used successfully. To measure the initiation of growth inhibition more accurately, seeds should be pregerminated (in darkness) in vermiculite moistened with 1 m*M* $CaSO_4$ until the roots and shoots are 2 to 4 mm in length. For oats, this pregermination period takes 2 days at ambient lab temperatures.

For measuring root growth, the pregerminated seedlings are washed to remove vermiculite and then placed on the surface of the moistened sand about 2 cm from the edge of the plate, with the roots pointing toward the distant edge. This permits the roots to grow about 5 cm prior to contacting the opposite edge. The petri dish covers are then placed over the seedlings and sealed in place with tape. The initial position of each root tip is marked on the petri dish covers (Figure 1), and then the petri dishes are placed in a dark incubator (25°C). To force root growth along the surface of the petri dish cover, they are placed (with roots pointing down) at an angle of about 20° from vertical. The new location of the roots is marked on the petri dish covers at timed intervals. At the completion of the experiment, the distances between the marks for each time interval are measured to the nearest 0.5 mm.

For measuring shoot growth the above procedure is used, except the seedlings are placed about 2 cm from the edge of the petri dish, with the shoot pointing toward the distant edge (Figure 1). Then the petri dishes are positioned in the incubator 20° from vertical so the shoots are forced to grow along the surface of the cover.

Table 1. Effect of Cinmethylin on Oat Root Growth

Herbicide Concentration (*M*)	Root Growth (mm)			
	0 to 6 hr	6 to 12 hr	12 to 18 hr	18 to 24 hr
0	5.0	5.7	7.5	7.1
1.0×10^{-8}	4.8	5.4	6.4	4.7
1.8×10^{-8}	4.7	4.3	2.8	1.8
3.3×10^{-8}	4.1	3.6	1.7	1.1
1.0×10^{-7}	3.9	3.1	1.5	0.9
1.0×10^{-6}	3.2	1.7	0.6	0.3

Source: El-Deek, H. and Hess, F. D., *Weed Sci.,* 34, 684, 1986. With permission.

RESULTS

The procedure described above is a very sensitive measure of growth inhibition in seedling plants. Data from a cinmethylin, exo-1-methyl-4-(1-methylethyl)-2-[(2-methyphenyl)methoxy]-oxabicyclo[2.2.1]heptane, experiment are shown in Table 1 for root growth and Table 2 for shoot growth (El-Deek and Hess, 1986). For cinmethylin, root growth inhibition occurred prior to shoot growth inhibition. Note that there is an interaction between herbicide concentration and the time to the first significant growth inhibition; i.e., higher herbicide concentrations inhibit growth in shorter time periods. Because this usually occurs, both time and concentration must be included as variables during the experiments.

COMMENTS

Herbicides with low water solubility can be added to the nutrient solution from a solvent stock solution. The final solvent concentration in the nutrient solution should be low (1% or less), and controls with and without the solvent must be included to determine the influence of the solvent on growth. Ethanol and dimethylsulfoxide are suitable solvents for many herbicides and at a concentration of 1% in the nutrient solution do not influence growth.

Prior to setting up the experiment, eight to ten holes should be made at the periphery of the petri dish covers with a hot needle. This will permit gas exchange during the experiment.

Table 2. Effect of Cinmethylin on Oat Shoot Growth

Herbicide Concentration (*M*)	Shoot Growth (mm)			
	0 to 12 hr	24 to 36 hr	48 to 60 hr	72 to 84 hr
0	3.3	8.9	7.2	5.2
1.0×10^{-8}	2.6	7.0	6.5	5.6
1.0×10^{-7}	2.6	8.3	5.6	3.5
1.0×10^{-6}	3.2	7.2	4.1	1.7
1.0×10^{-5}	2.6	6.1	1.7	0.3

Source: El-Deek, H. and Hess, F. D., *Weed Sci.,* 34, 684, 1986. With permission.

Occasionally, forces associated with root and shoot growth cause the seed to move from its original position, which can yield inaccurate measurements. Thus, the initial location of the seed should be marked on the petri dish cover (Figure 1). If a seed moves during the growth period, data from that plant should be discarded.

Measurements can be reported as growth per time period or accumulative growth achieved during each time period. Data can be reported in tables or three-dimensional graphs. To provide adequate data for statistical analysis, four to six plants are included in each petri dish (subsamples) with four to six petri dishes per treatment (replicates). The entire experiment should be repeated to confirm the results.

REFERENCES

El-Deek, H. and Hess, F. D., Inhibited mitotic entry is the cause of growth inhibition by cinmethylin. *Weed Sci.,* 34, 684, 1986.

Hess, F. D., Herbicide interference with cell division in plants, in *Target Sites of Herbicide Action,* Böger, P. and Sandmann, G., Eds., CRC Press, Boca Raton, FL, 1989, p. 85.

Hess, F. D., Determining causes and categorizing types of growth inhibition induced by herbicides, in *Biochemical Responses Induced by Herbicides,* Moreland, D. E., St. John, J. B., and Hess, F. D., Eds., Symp. Ser. No. 181, American Chemical Society, Washington, D.C., 1982, p. 207.

Hoagland, D. R. and Arnon, D. I., The water culture method for growing plants without soil, *Univ. Calif. Agric. Exp. Stn. Circ.,* Vol. 347, 1950.

CHAPTER 26

Analysis of the Mitotic Index in Root Tip Meristems as a Tool to Determine Herbicide Cell Cycle Specificity

Thomas L. Rost and F. Dan Hess

INTRODUCTION

The vascular plant body has a bipolar axis terminated by a root and shoot apical meristem. These meristems are the locations of cycling cells and the sites where developmental decisions are made which determine the identities of specific cell types and tissues. Herbicides which are developmentally active tend to inhibit one or more of the processes which are localized in meristems. Exactly which event is the primary target for herbicide action is often difficult to determine. In some instances the herbicide inhibits growth directly as a primary response; in other instances growth is stopped, but as a secondary effect. This chapter will discuss the use of a simple experiment designed to help determine if herbicide-induced growth inhibition affects cell cycle progression directly or indirectly.

Herbicides generally affect meristem cell behavior by one of three patterns: (1) preprophase inhibition, (2) spindle inhibition, or (3) cytokinesis inhibition (Rost, 1977).

Preprophase inhibition means that cell cycle progression is blocked in G_1, S, or G_2, before the cell initiates prophase. An example is the herbicide

0-87371-539-X/93/$0.00 + $.50

chlorsulfuron (Rost, 1984). In this instance, chlorsulfuron inhibits the activity of acetolactate synthetase, which blocks the biosynthesis of branched amino acids, causing the buildup of toxic intermediates and selectively stopping cell cycle progression within a few hours (Rost et al., 1990). Cells are arrested in G_1 and G_2 and can be reversed by adding branched amino acids to the growth medium.

Examples of spindle inhibitors are the plant alkaloid colchicine and herbicides like trifluralin (Hess and Bayer, 1974). These herbicides block the polymerization of tubulin into microtubules, thus disrupting chromosome movement during mitosis. Cells treated with these chemicals usually continue enlarging, but become rounded instead of elongate, and root tips exhibit terminal swelling (see Chapter 28).

Cytokinesis inhibition is induced by chemicals, such as caffeine, which inhibit cell plate formation. Cells treated this way become binucleate. Several researchers have used this inhibitor as an experimental tool to study many aspects of cell cycle behavior (Rost, 1977).

PROCEDURES

Treated root tips (0.5 cm in length) are collected and fixed in absolute alcohol:acetic acid (3:1) for at least 20 min. They are then hydrolyzed in 5 *N* HCl for 10–20 min, rinsed in water, and stained with basic fuchsin by the Feulgen reaction (Jensen, 1962). The hydrolysis step forms aldehyde groups on the DNA molecule, which then binds with the dye. With this method the chromosomes are stained deep red. In some unusual plants (e.g., *Azolla* fern roots) the DNA does not stain well by the Feulgen reaction. In those cases it may be necessary to use a fluorescent DNA dye, such as DAPI (4′, 6-diamidino-2-phenylindole dihydrochloride) (Levi et al., 1986), which binds directly to the DNA bases. This is an easier procedure to use, but the cells must be examined with fluorescent optics. Fluorescent dyes are also a little more difficult to use for cell counts because they fade quite rapidly while being exposed to light.

After staining, individual root tips (1 mm) are placed on a microscope slide and squashed. This is accomplished by adding a small drop of 45% acetic acid, drawing most of the acid off with a tissue, and then squashing the root tips with a glass rod using an up and down motion. When the root tip is well squashed, add a little more acid and smear the cells around gently so that they are spread in a circle about 1 cm in diameter. This will ensure mixing of the cells and will help to eliminate clumping.

Some roots (e.g., most grasses) will not squash easily using the method described above. In those cases it will be necessary to loosen the cells from each other by enzyme digestion. A common method is to place the roots in water either immediately after fixation or after staining, followed by softening for 5 hr at 27°C in a 1% solution of pectinase in 0.1 *M* acetate buffer, pH 4.0. After this the roots will usually squash very easily.

Slides are made permanent by freezing them on a block of dry ice, then popping off the cover glasses and placing the slides in 100% ethanol. New cover glasses can then be mounted on the slides with Euparal, which is soluble in ethanol, or the slides can be transferred to xylene or a xylene substitute (e.g., Hemo De from Fisher) and cover glasses then mounted.

Mitotic index counts are made by scoring slides for the number of mitotic figures present per population counted and expressing that as a percent. Usually a minimum of 500 to 1000 cells are counted per slide, with four to five slides used per data point. It is often useful to count the number of cells in each mitotic phase. This can be important because it shows if the chemical specifically affects one phase of mitosis.

RESULTS

Mitotic index data from the use of three different herbicides (chlorsulfuron, napropamide, and oryzalin) will be used to demonstrate this procedure.

Chlorsulfuron, as mentioned earlier, inhibits branched amino acid biosynthesis and blocks cell cycle progression at very low concentrations (28 n*M*). Pea roots were excised and cultured for 24 hr; then they were treated with chlorsulfuron and sampled at intervals over a 24-hr period (Rost and Reynolds, 1985). Roots were fixed, stained, and prepared as discussed above. The mitotic index curve showed that the number of cells in mitosis declined very rapidly by 4 hr and was essentially 0 by 12 hr (Figure 1). The shape of this curve and the fact that such a low dose of herbicide stops cell cycle progression indicate that the mode of action of this chemical directly affects mitotic progression.

The mode of action of napropamide is not known, but is thought to involve the inhibition of RNA and/or protein synthesis. Pea roots were grown as described above and then treated with 20 μ*M* napropamide. The mitotic index curve showed a much more gradual inhibition response with no reduction in the number of mitotic figures until after 6 to 8 hr of treatment (Figure 1). This observation is evidence in support of the conclusion that the mode of action of this chemical is not directly related to cell cycle inhibition (DiTomaso et al., 1988).

Oryzalin is an herbicide in the dinitroaniline class which is known to inhibit the polymerization of microtubules. These herbicides act like colchicine by blocking mitosis in early metaphase. This chemical does not block the entry of cells into mitosis and instead slows down their exit. In addition, the affected cells do not separate their chromosomes and will become tetraploid. A mitotic index curve for cultured *Convolvulus* roots treated with 10^{-7} *M* oryzalin is shown in Figure 1 (Rost and Bayer, unpublished data). Note that in this kind of response, since the exit out of mitosis is slowed, the curve shows an accumulation response.

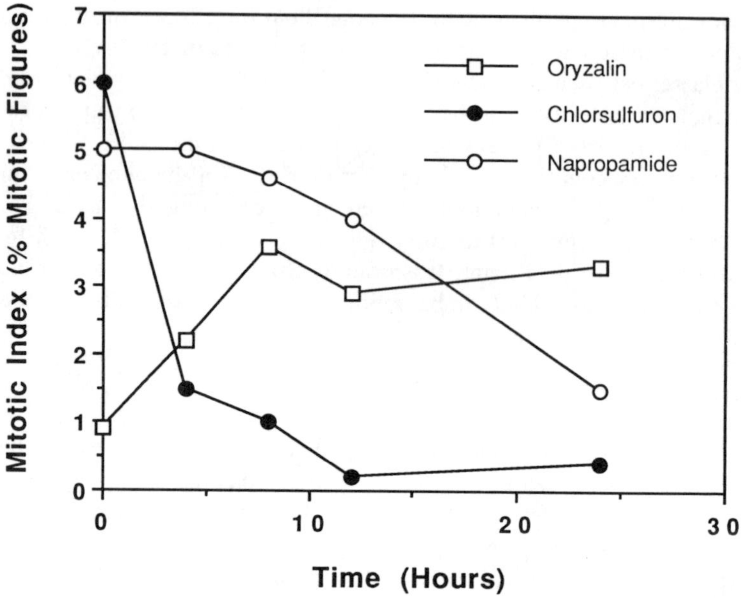

Figure 1. Mitotic index (MI) curves. Oryzalin is a microtubule polymerization inhibitor and shows an accumulation-type curve. Data from cultured *Convolvulus arvense* roots treated with 10^{-7} *M* oryzalin. Chlorsulfuron shows a direct inhibition response as indicated by the immediate fall in the MI curve after using a very low dose of herbicide (20 n*M*). This curve is taken from Rost and Reynolds (1985) from cultured pea (*Pisum sativum* cv Alaska) roots. Napropamide is an indirect mitotic inhibitor, so it shows a gradual MI inhibition curve. Thses data are from cultured pea roots treated with 20 μ*M* napropamide (DiTomaso et al., 1988). Oryzalin, 4-(dipropylamino)-3,5-dinitrobenzenesulfonamide; chlorsulfuron, 2-chloro-*N*-[[(4-methoxy-6-methyl-1,3,5-triazin-2-yl)amino]carbonyl]benzenesulfonamide; napropamide, *N,N*-diethyl-2-(1-naphthalenyloxy)propanamide.

COMMENTS

A common problem in using root tips for herbicide effect experiments is the variability between roots. It is important in doing these experiments to select roots of similar length and appearance and to use fresh seed batches. In addition, staining responses to the Feulgen reaction are sometimes different between plant species, and it will be necessary to try several different hydrolysis times when this method is attempted for the first time.

The shape of the mitotic index curve can be used to initially characterize whether an herbicide is a direct cell cycle inhibitor. An immediate reduction in mitotic index with a low dose of chemical is evidence that a direct effect is involved; chlorsulfuron is an example. A gradual inhibition response, like with napropamide, is an example of an herbicide which inhibits the cell cycle as a secondary response. A third response, an accumulation curve, occurs when roots are treated with microtubule polymerization inhibitors.

REFERENCES

DiTomaso, J. M., Rost, T. L. and Ashton, F. M., The comparative cell cycle and metabolic effects of the herbicide napropamide on root tip meristems. *Pest. Biochem. Physiol.,* 31, 166, 1988.

Hess, D. and Bayer, D., The effect trifluralin on the ultrastructure of dividing cells of the root meristem of cotton (*Gossypium hirsutum* L. 'Acala 4-42'). *J. Cell Sci.,* 15, 429, 1974.

Jensen, W. A., *Botanical Histochemistry, Principles and Practices.* W. H. Freeman, San Franciso, 1962.

Levi, M., Tarquini, F., Sgorbati, S. and Sparvoli, E. Determination of DNA content by static cytofluorimetry in nuclei released from fixed plant tissue. *Protoplasma,* 132, 64, 1986.

Rost, T. L., Responses of the plant cell cycle to stress, in *Mechanisms and Control of Cell Division,* Rost, T. L. and Gifford, E. M., Jr., Eds., Dowden, Hutchison and Ross, Stroudsburg, PA, 1977, p. 111.

Rost, T. L., The comparative cell cycle and metabolic effects of chemical treatments on root tip meristems. III. Chlorsulfuron. *J. Plant Growth Regul.,* 3, 51, 1984.

Rost, T. L., Gladish, D., Steffen, J. and Robbins, J., Is there a relationship between branched amino acid pool size and cell cycle inhibition in root tips after treatment with imidazolinone herbicides? *J. Plant Growth Regul.,* 9, 227, 1990.

Rost, T. L. and Reynolds, T., Reversal of chlorsulfuron-induced inhibition of mitotic entry by isoleucine and valine. *Plant Physiol.,* 77, 481, 1985.

CHAPTER 27

Estimates of Cell Cycle Duration in Meristems of Herbicide-Treated Plants

Thomas L. Rost and F. Dan Hess

INTRODUCTION

Plant meristems are the sources of cells making up the plant body and are the sites for the initiation of cell elongation and cell differentiation. Plant form and the orderly development of plant organs are at least in part regulated by the rate of cell division in the root and shoot apical meristems. Herbicides which inhibit growth often act directly or indirectly on cells by interfering with one or more steps involved in the control of the rate and timing of the cell cycle.

Two different methods are most commonly used by plant biologists to determine the duration of the cell cycle in meristems (for reviews see Gould, 1984a, and Webster and MacLeod, 1980). The colchicine method involves treating tissues with the alkaloid colchicine to inhibit the progression of cells through mitosis. Estimates of cell cycle times are determined by calculations based on the slope of the mitotic accumulation curve (Van't Hof, 1968). This method has the advantage of being able to estimate cell cycle times for specific tissues, but has the serious disadvantage of involving a chemical agent which inhibits microtubule polymerization and may compound the inhibitory effects of the herbicides being studied. The second and more useful method was developed by Quastler and Sherman (1959) and involves the use of

0-87371-539-X/93/$0.00 + $.50

^{3}H-thymidine incorporation into cells undergoing DNA synthesis. After incorporation for a short time period, cells are allowed to enter into G_2 and mitosis. The cells which are labeled with ^{3}H-thymidine will now be tagged as they progress through the cell cycle. The kinetics of this progression curve can be used to estimate the duration of cell cycle stages and the length of the cell cycle, or cell cycle time (Gould, 1984b; Van't Hof, 1968). This chapter will discuss the Quastler/Sherman or fraction of labeled mitoses (FLM) method to measure the duration of the cell cycle in apical meristems.

An analysis of the average duration of the cell cycle may be important in determining the mode of action of an herbicide. Metabolic inhibitors may simply slow cell cycle progression, while those which act directly on the mitotic apparatus will stop the cycle. Herbicides which affect a specific cell cycle stage — for example, DNA synthesis — change the duration of S phase and in this way slow the duration of the cell cycle. This method is a useful way to isolate factors which affect the cell cycle and its phases in plant meristems.

PROCEDURE

The following assumptions must be made before using this procedure:

1. The cells in the meristem are asynchronous.
2. They are equally distributed throughout the cell cycle.
3. They are in steady state; that is, the same number of cells are entering and leaving the cycle.

Intact plants, cultured organs, or tissue cultures may be used for this method.

Plant organs or cultures are treated with ^{3}H-thymidine (usually for 10 to 30 min, 0.1 μCi/mL) by transferring the tissues to the isotope solution or adding the isotope into the culture medium.* After the treatment time a nonradioactive chase of about 50 μ*M* thymidine is sometimes used to stop ^{3}H-thymidine incorporation into DNA. The tissues are then transferred to fresh medium and allowed to grow.

Samples are usually taken at 2-hr intervals and placed in 3:1 fixative (absolute ethanol:acetic acid) for several hours. The samples are stained with the Feulgen reaction (Jensen, 1962; see also Chapter 26 in this volume). Tissues are then squashed on microscope slides, made permanent by freezing on a block of dry ice, plunged into 100% ethanol, and then air-dried. Slides are coated with liquid photographic emulsion (e.g., Kodak NTB3) in a darkroom using a no. 1 safelight. Be sure to read the instructions for using the specific emulsion, and pay particular attention to the recommended safelight conditions before opening the box. Usually, the emulsion is placed into a

* Special procedures for handling and disposal of radioactive materials must be used. Consult with your health and safety department or committee for insturctions.

40-mL beaker and the beaker heated to 45°C in a shallow water bath. The emulsion takes 10 to 20 min to melt. The emulsion can then be poured into a narrow container (plastic mailers for single microscope slides work well) so that the slides can be dipped one at a time into the emulsion. Slides are dried by standing them up in a rack or placing one end on a paper towel and leaning the other end against a vertical surface. After a 1- to 2-hr dark drying period, the slides are placed in small slide boxes which contain a small gauze sack filled with desiccant (e.g., $CaSO_4$). The slide boxes are then sealed with electrical tape, wrapped in aluminum foil, and placed in a refrigerator approved for radioactive storage. The exposure time in the refrigerator varies, but is usually 1 to 3 days. The slides are then developed (e.g., DK19 developer) and fixed in a darkroom under a no. 1 safelight, washed in running water for 10 to 20 min, dehydrated in an ethanol series, and a cover glass is then mounted in Euparal or some other mounting medium.

The autoradiograph slides are then scored for the percentage of total mitotic figures (mitotic index, MI) and of labeled mitotic figures (or FLM). The total number of cells, interphase and mitotic, which are labeled should also be counted. For most experiments, a labeled cell is any cell where more than five silver grains above the backbround count are found above the nuclei or chromosomes. The percentage of labeled cells is called the labeling index (LI), and after a lag time of several hours it should remain constant. This LI percentage is the growth fraction (GF), which is an estimate of the number of cells in the meristem that are actually cycling. In most plants a GF of 70 to 90% is normal.

RESULTS

The graph of percent labeled mitotic figures vs. time in a hypothetical experiment is shown in Figure 1. Several bits of kinetic information can be determined directly from the curve. Halfway up the ascending part of the first curve is an estimate of the time of G_2 plus one half the time of M ($T_{G2+1/2M}$). An estimate of the duration of DNA synthesis (T_S) is the time marked by a line halfway up and across the first bell-shaped curve (Figure 1). These values are quite trustworthy in most experiments.

The total cell cycle time (TC) is determined from the time delimited from the peaks of the first and second curves. All that is needed now are estimates of the durations of mitosis (T_M) and G_1 (T_{G1}). One method used to estimate T_M is based on the assumption that the time it takes to traverse a phase is proportional to the number of cells in that phase (Gould, 1984a; Van't Hof, 1968). The time for mitosis can be estimated by multiplying the percentage of cells in mitosis (e.g., 10%) times the estimate of the cell cycle time (e.g., 14 hr). In this example the time of mitosis would be 10% of 14 hr, or 1.4 hr.

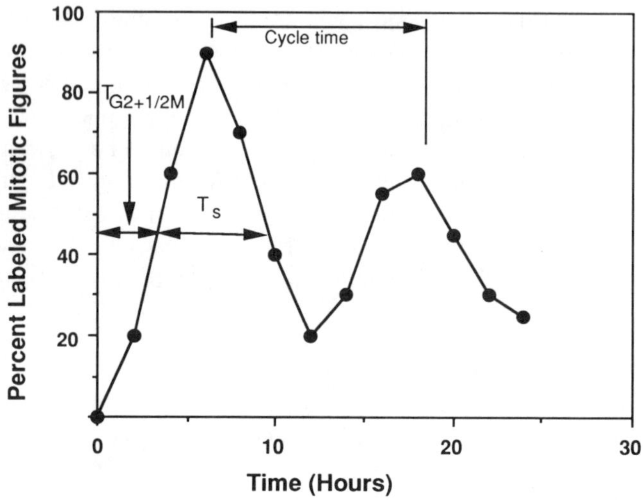

Figure 1. Hypothetical graph of percent labeled mitotic figures plotted against time after the addition of ^{3}H-thymidine. An estimate of the cell cycle duration (cycle time) is the time from the peak of the first curve to the peak of the second. The time of G_2 plus one half of mitosis ($T_{G2+1/2M}$) is equal to the time from the addition of thymidine to one half the distance up the first curve. The time of DNA synthesis (T_S) is equal to the time between the halfway points of the first curve.

The duration of G_2 can be calculated once the time of M has been estimated ($G_2 = G_{2+1/2M} - 0.5T_M$). The duration of G_1 is determined by subtraction as follows:

$$T_{G1} = TC - T_{G2} - T_S - T_M$$

Estimates of cell cycle times for two plants are shown in Table 1.

COMMENTS

There are problems inherent to this method. One is that these cell cycle parameter estimates are averages of all of the cells from the meristem. This is partially in error because cells in different parts of a meristem cycle at different rates (Evans and Van't Hof, 1975). The average cycle time, on the other hand, is a useful measure of the overall metabolic condition of the

Table 1. Cell Cycle Durations[a]

Plant	TC	T_{G2}	T_M	T_{G1}	T_S
Pisum sativum	14.0	3.0	1.2	5.0	4.5
Helianthus annuus	7.8	1.5	0.6	1.2	4.5

[a] Values are in hours.
Source: Data from Van't Hof (1974).

meristem. In addition, the herbicide may have such a severe inhibition effect that the second labeled mitotic wave may not appear. Even though the cell cycle has been terminated, the durations of $G_{2+1/2M}$ and S can provide useful information.

REFERENCES

Evans, L. S. and Van't Hof, J., The age-distribution of cell cycle populations in plant root meristems. *Exp. Cell Res.,* 90, 401, 1975.

Gould, A. R., Cell cycle analysis by convential methods, in *Cell Culture and Somatic Cell Genetics of Plants,* Vol. 1, Vasil, I. K., Ed., Academic Press, New York, 1984a, p. 753.

Gould, A. R., Control of the cell cycle in cultured plant cells. *Crit. Rev. Plant Sci.,* 1, 315, 1984b.

Jensen, W. A., *Botanical Histochemistry, Principles and Practice,* W. H. Freeman, San Francisco, 1962.

Quastler, H. and Sherman, F. G., Cell population kinetics in the intestinal epithelium of the mouse. *Exp. Cell Res.,* 17, 420, 1959.

Van't Hof, J., Experimental procedures for measuring cell population kinetic parameters in plant root meristems, in *Methods in Cell Physiology,* Vol. 3, Prescott, D. M., Ed., Academic Press, New York, 1968, p. 95.

Van't Hof, J., The duration of chromosomal DNA synthesis, of the mitotic cycle, and of meiosis of higher plants, in *Handbook of Genetics,* Vol. 2, King, R. C., Ed., Plenum Press, New York, 1974, p. 363.

Webster, P. L. and MacLeod, R. D., Characteristics of root apical meristem cell population kinetics: a review of analyses and concepts. *Environ. Exp. Bot.,* 20, 335, 1980.

CHAPTER 28

Analysis of Cell Elongation as a Tool to Determine Herbicide Effects on Plant Growth

F. Dan Hess and Thomas L. Rost

INTRODUCTION

Cell enlargement is an essential part of growth in plant meristems. In a cell with a wall capable of extension (yielding), enlargement occurs when hydrostatic pressure inside the cell exerts a force against the wall. In many cells, enlargement is preferentially in length (anisotropic). Anisotropic growth is due to different locations on the cell wall having dissimilar yielding potentials caused by different orientations of the cellulose microfibrils in the wall. For example, prior to enlargement, microfibrils in the longitudinal walls of root meristem cells are arranged transversely. This microfibril orientation gives the wall strength in the transverse direction; thus, wall yielding (cell enlargement) occurs preferentially in the longitudinal direction and the cells enlarge into cylinders rather than spheres. During enlargment new microfibrils are deposited in a transverse orientation on the inside of the wall, which continues to provide wall strength in this direction.

Herbicides can either inhibit or disrupt cell enlargement (see Hess, 1982, for review). Inhibition of cell enlargement can be caused by membrane disruption (e.g., diphenyl ethers and bipyridyliums), metabolic changes within the cell (e.g., chloroacetamides), or alterations in a process (e.g., hydrogen ion extrusion) necessary for cell enlargement (e.g., aryloxyphenoxypropion-

0-87371-539-X/93/$0.00 + $.50

ates). Herbicides that disrupt membranes do inhibit growth, but this is of little importance in their overall mechanisms of action.

One mechanism whereby herbicides disrupt cell enlargement is by inducing wall loosening. Auxin agonist herbicides (e.g., 2,4-D) disrupt cell enlargement by this mechanism. As discussed in the first paragraph, specific microfibril orientation is essential for normal enlargement in many cells. Microtubules are thought to be responsible for the orientation of microfibrils (see Hepler and Palevitz, 1974, for review); therefore, herbicides that inhibit microtubule polymerization (e.g., dinitroanilines and pronamide) or micrortubule function (e.g., phenylcarbamates and DCPA, [dimethyl 2,3,5,6-tetrachloro-1,4-benzenedicarboxylate]) cause aberrant cell enlargement. This results in spherical cells (isotropic cell enlargement) and swollen apical regions. Inhibition of cellulose biosynthesis by herbicides (e.g., dichlobenil and isoxaben) also causes aberrant cell enlargement and subsequent swelling of apical regions of germinating seedlings because new microfibrils are not being deposited along the inside of the cell wall during cell enlargement.

PROCEDURES

Cell enlargement in germinating seedlings can be assessed qualitatively by microscopic examination of longitudinally sectioned root and shoot tips after standard fixation, dehydration, paraffin embedding, sectioning, and staining procedures (see Jensen, 1962, for detailed procedures). However, to specifically assess quantitative cell enlargment, the processes of cell division and cell enlargement must be separated. This can be accomplished by assessing growth in specific tissues known to be undergoing only cell enlargement. The oat (*Avena sativa*) coleoptile straight growth test described by Bonner (1933) and the pea (*Pisum sativum*) hypocytyl straight growth test described by Thimann and Schneider (1939) remain useful techniques. An alternative is to assess single cells in model plant systems (e.g., algal cells).

The method most commonly used to quantify cell enlargment is the oat coleoptile assay as refined by Nitsch and Nitsch (1956). Prior to using this assay it must be confirmed that the herbicide being assessed is active on oats. To obtain suitable tissue for analysis, oat seeds (florets, botanically) are planted in vermiculite moistened with 1 m*M* $CaSO_4$ and then germinated and grown in a dark incubator at 25°C for 5 days. At this time the coleoptiles should be 3 to 4 cm in length. To prevent phytochrome growth responses, all manipulations of the seedlings described below must be conducted in a room illuminated only by a green safelight.

Herbicide treatments are prepared by diluting herbicide stock solutions with 10 m*M* phosphate buffer (pH 5.2). If solvents are needed for dissolving the herbicide, the final concentration of the solvent should be low (1.0% or less), and controls with and without solvent must be included in the experiment. Then, 10 mL of the treatment solution is placed in 9-cm round petri dishes. Coleoptile sections (5 mm) are cut from the oat seedlings starting 3 mm below the tip. Subsequent growth in this section is due almost entirely to cell enlargement (Nitsch and Nitsch, 1956). Ten coleoptiles are then floated on the treatment solution in each petri dish; covers are placed over the dishes, and they are incubated on a shaker (one oscillation per 1 to 2 sec; see Comments section) at 25°C in the dark for 24 hr. At the end of the incubation, the coleoptiles are measured to the nearest 0.1 mm using a binocular microscope equipped with a calibrated ocular micrometer. During the data analysis, individual coleoptiles should be considered subsamples. To provide adequate data for statistical analysis, three to four petri dishes per treatment are necessary, and the experiment should be repeated to confirm the results.

For herbicides that are more active on dicotyledonous plants than on monocotyledonous plants, the above procedure can be adapted to peas. Peas (cv. Alaska) are planted in vermiculite moistened with 1 m*M* $CaSO_4$ and then germinated and grown in the dark at 25°C until the third internode of the hypocotyl is developed enough for 10-mm sections to be cut just below the hypocotyl hook. To achieve this growth stage requires about 10 days. As in the oat coleoptile test, all tissue manipulations should be carried out in a dark room illuminated with a green safelight. As for the coleoptile test, the treatments are prepared in 10 m*M* phosphate buffer. Sections, 10 m*M* in length, are cut from the third internode just below the hypocotyl hook. The number of sections per petri dish, the number of petri dishes per treatment, incubation conditions, and hypocotyl measurement are the same as in the oat coleoptile test.

The most suitable algae to use in cell elongation tests are *Chara* and *Nitella*. After cell division, single internode cells (20 μm or so in length) will undergo anisotopic cell enlargement until reaching a final length of several centimeters (Figure 1).

An aseptic culture of *Nitella* or *Chara* can be initiated from oospores using the procedure of Imahori and Iwasa (1965). Cultures are maintained in autoclaved soil water medium* and experiments conducted using one of several published artificial media (e.g., Green, 1968).

One apical algal segment, 2 to 3 cm in length, is transferred to a petri dish (10 cm diameter × 2 cm deep) containing 40 mL of treatment solution. The internode cells to be measured should be 300 μm or less in length at the onset of the experiment. The treatment solutions are prepared the same as for the oat coleoptile study described above. Cell enlargement (length and diameter) of the selected internode is measured *in situ* each 24-hr period using a binocular microscope equipped with a calibrated ocular micrometer. Data collected from at least five petri dishes are needed for statistical data analysis.

* Soil water medium: 1 L water, 100 g sandy loam field soil, 15 mg $CaCO_3$. This mixture is steamed in an autoclave for 1.5 hr on 3 consecutive days, then autoclaved for 15 min on the final day. The mixture is allowed to settle for 2 to 4 weeks prior to use.

Figure 1. Internode cell enlargement in the alga *Chara.* The two micrographs are from the same internode cell. The inset was recorded 10 days prior to the main micrograph. Magnification = 10X. (Reprinted with permission from Hess, F. D., in *Biochemical Responses Induced by Herbicides,* Moreland, D. E., St. John, J. B., and Hess, F. D., Eds., ACS Symp. Ser. 181, American Chemical Society, Washington, D.C., 1982, p. 207. Copyright 1982 American Chemical Society.)

RESULTS

Herbicides that inhibit cell enlargement can be assessed efficiently in the oat and pea straight growth tests. The results are usually reported as the amount of elongation that occurs over the 24-hr incubation time. The effect of cinmethylin on oat coleoptile cell elongation with and without added indole-3-acetic acid (IAA) is shown in Table 1.

For herbicides that are suspected of disrupting cell enlargement, the algal test is most suitable for obtaining quantitative data. In this test a single cell can be studied; thus, there is no effect of adjacent cells on the cell being measured. The influence of trifluralin (2,6-dinitro-*N*,*N*-dipropyl-4-(trifluoromethyl)benzeneamine) on cell enlargement in *Chara* internode cells is shown in Figures 2 and 3. Data can also be expressed as relative growth rate:

$$[(\ln L_2 - \ln L_1)/(t_2 - t_1)]$$

where L = length and t = time.

COMMENTS

For the oat coleoptile and pea hypocotyl tests, several tissue sections can be cut at one time with two razor blades spaced the correct distance apart by a rigid plastic spacer.

Gentle shaking of the oat/pea sections during incubation is very important. Constant movement will ensure that the tissue sections remain straight, which permits accurate measurements. If some tissue segments curve during the 24-hr incubation period, increasing the shaking speed may reduce the amount of curvature. However, if shaking is too vigorous, tissue damage will occur. If increased shaking does not provide straight tissue segments, additional tissue pieces can be added to each petri dish and then the ten straightest sections measured.

IAA (50 μg/L) can be added to the oat coleoptile and pea hypocotyl test to determine the influence of the herbicide on auxin-induced cell enlargement.

For the *Nitella* and *Chara* cell elongation procedure, it is necessary to assume that the mechanism of cell enlargement is the same as in higher plants. In these algae there is evidence that cell wall microfibrils are oriented by microtubules and that transverse microfibril orientation is the cause of anisotropic cell enlargement (Wasteneys and Williamson, 1987).

Table 1. The Effect of Cinmethylin[a] on Cell Enlargement in Oat Coleoptile Segments with and without Indole-3-acetic Acid (IAA)

Herbicide Concentration (*M*)	Coleoptile Length (mm)	
	With IAA	Without IAA
0	5.8	2.7
1×10^{-6}	5.6	2.8
1×10^{-5}	5.9	2.5
1×10^{-4}	4.2	1.9

Source: El-Deek, M. H. and Hess, F. D., *Weed Sci.*, 34, 684, 1986. With permission.

[a] Cinmethylin is exo-1-methyl-4-(1-methylethyl)-2-[(2-methylphenyl)methoxy]-7-oxabicyclo[2.2.1]heptane.

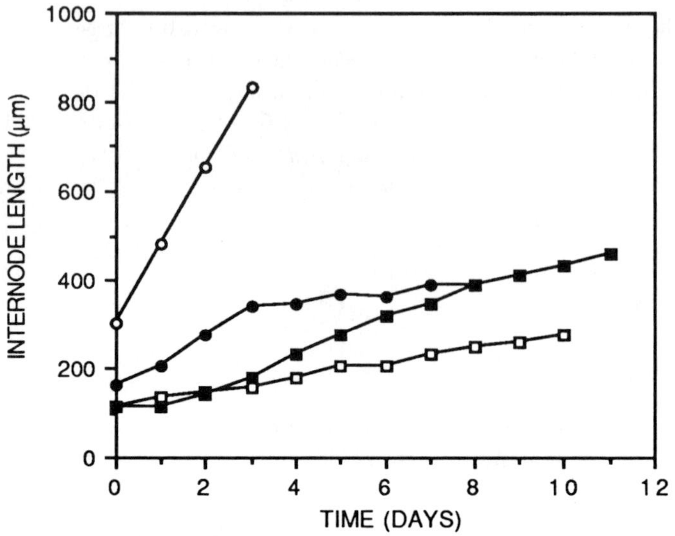

Figure 2. *Chara* cell elongation in a nontreated internode and a trifluralin (1 μ*M*), 2,6-dinitro-*N,N*-dipropyl-4-(trifluoromethyl)benzeneamine, treated internode. Circles = cell length; squares = cell diameter; open = control; closed = treated.

Figure 3. Apex of *Chara* treated with 1 μ*M* trifluralin for 9 days. Magnification = 12X. (Reprinted with permission from Hess, F. D., in *Biochemical Responses Induced by Herbicides,* Moreland, D. E., St. John, J. B., and Hess, F. D., ACS Symp. Ser. 181, American Chemical Society, Washington, D.C., 1982, p. 207. Copyright 1982 American Chemical Society.)

REFERENCES

Bonner, J., The action of the plant growth hormone. *J. Gen. Physiol.*, 17, 63, 1933.

El-Deek, H. and Hess, F. D., Inhibited mitotic entry is the cause of growth inhibition by cinmethylin. *Weed Sci.*, 34, 684, 1986.

Green, P. B., Growth physics in *Nitella:* a method for continuous *in vivo* analysis of extensibility based on a micro-manometer technique for turgor pressure. *Plant Physiol.*, 43, 1169, 1968.

Hepler, P. K. and Palevitz, B. A., Microtubules and microfilaments. *Annu. Rev. Plant Physiol.*, 25, 309, 1974.

Hess, F. D., Determining causes and categorizing types of growth inhibition induced by herbicides, in *Biochemical Responses Induced by Herbicides,* Moreland, D. E., St. John, J. B., and Hess, F. D., Eds., Symp. Ser. No. 181, American Chemical Society, Washington, D.C., 1982, p. 207.

Imahori, K. and Iwasa, K., Pure culture and chemical regulation of the growth of Charophytes. *Phycologia,* 4, 127, 1965.

Jensen, W. A., *Botanical Histochemistry, Principles and Practice,* W. H. Freeman, San Francisco, 1962.

Nitsch, J. P. and Nitsch, C., Studies on the growth of coleoptile and first internode sections. A new, sensitive, straight-growth test for auxins. *Plant Physiol.*, 31, 94, 1956.

Thimann, K. V. and Schneider, C. L., The relative activities of different auxins. *Am. J. Bot.*, 26, 328, 1939.

Wasteneys, G. O. and Williamson, R. E., Microtubule orientation in developing internodal cells of *Nitella:* a quantitative analysis. *Eur. J. Cell Biol.*, 43, 14, 1987.

CHAPTER 29

Immunofluorescence Microscopy of Tubulin to Determine Phytotoxin Effects on the Plant Cytoskeleton*

Timothy D. Sherman and Kevin C. Vaughn

INTRODUCTION

The use of fluorescence microscopy to determine the distribution of microtubules in plant cells has led to a renaissance of work on the plant cytoskeleton. Previously, determination of the three-dimensional arrangement of microtubules required the preparation of serial section electron micrographs and reconstruction of the arrangement from these images. Now, with a relatively simple protocol (Wick et al., 1981), the three-dimensional distribution of microtubules and the effect of phytotoxins on these arrays may be determined in a single preparation. This chapter outlines a protocol (Lehnen et al., 1990) that works well for many plant species and gives examples of the use of this technology in elucidating phytotoxin mode of action.

PROCEDURE

1. Onion seeds are placed in 9-cm petri dishes on Whatman 1 filter paper moistened with H_2O. The seedlings are allowed to germinate and grow

for 3 to 5 days in the dark before treatment with herbicide solutions or a control solution containing only the solvent used to dissolve the herbicide (normally $\leq$1% (v/v) methanol or acetone). The seedlings remain in the treatment solution for 2 to 24 hr in the dark.

2. Root tips are fixed with 4% (w/v) paraformaldehyde in 50 mM PIPES (pH 6.9)/5 m*M* EGTA/5 m*M* $MgSO_4$/10% (v/v) DMSO for 1 hr. All steps of fixation and processing are carried out at room temperature.
3. Rinse in three exchanges, 10 min each, in 50 m*M* PIPES (pH 6.9)/5 m*M* EGTA/5 m*M* $MgSO_4$.
4. Digest root tips in 1% (w/v) Cellulysin (Calbiochem, LaJolla, CA) in 50 m*M* PIPES (pH 6.9)/5 m*M* $MgSO_4$/100 μ*M* $CaCl_2$/1% (w/v) bovine serum albumin (BSA) for 8 min.
5. Rinse in three exchanges, 10 min each, in 50 m*M* PIPES (pH 6.9)/5 m*M* $MgSO_4$/100 μ*M* $CaCl_2$.
6. Remove unpolymerized tubulin and improve antibody penetration by incubating in rinse buffer containing 0.1% (v/v) Triton X-100.
7. Macerate the terminal 1–2 mm of root in 1% (w/v) BSA in 20 m*M* $NaPO_4$ (pH 7.2)/150 m*M* NaCl (PBS-BSA) on a microscope slide with a glass pestle made from a flame-polished stirring rod.
8. Cover the cells with a coverslip and squash the preparation gently between layers of paper towel.
9. Separate the slide and the coverslip. Cells that adhere to the coverslip as well as those on the slide may be processed. Incubate the cells in a 1:100 dilution of monoclonal anti-chick brain tubulin (Amersham International, Amersham, U.K.) for 40 min in a moist chamber.
10. Rinse the cells four times with PBS (5 min per rinse).
11. Incubate the cells in a 1:100 dilution of fluorescein isothiocyanate (FITC)-conjugated goat anti-mouse IgG in PBS-BSA for 40 min.
12. Rinse the cells as in step 10.
13. Mount the cells in PBS/50% (v/v) glycerol, containing 1.5% (w/v) propyl gallate to retard FITC fading.
14. View with a microscope equipped with epifluorescence optics and filter set combinations appropriate for FITC.
15. Photograph the images using Kodak Tri-X pan film at an ASA of 1600.

RESULTS

Tubulin immunofluorescence is a relatively simple and effective method for determining whether a particular compound affects microtubule arrays (Vaughn and Lehnen, 1991). For example, the herbicides dithiopyr and pronamide cause the loss of all of the microtubules except for a tuft that remains at the kinetochore. The herbicide terbutol causes star anaphase configurations that result from the microtubules aggregating around a small area rather than spreading away from the two poles of the cell. DCPA affects the phragmoplast arrays by causing both a fragmentation and misorientation of the normal phragmoplast array. Examples of control and barban-treated onion root tip cells processed for tubulin immunofluorescence are shown in Figure 1. In the barban-treated cells, the spindles are focused into many poles rather than

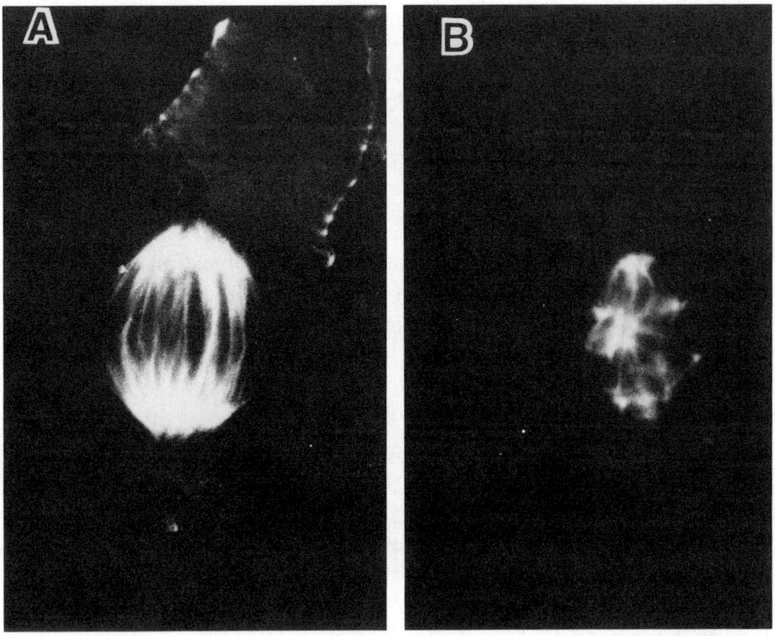

Figure 1. Immunofluorescence micrographs of (A) control and (B) barban-treated onion root tips. The spindle microtubule array in the control exhibits the characteristic bipolar spindle, whereas the barban-treated cell has a multipolar spindle. (Magnification 400X.)

exhibiting the bipolar spindle characteristic of the control cells. None of these determinations of herbicide mechanism could have been accomplished as quickly or as definitively with the more classical methods of analysis. Besides the analysis of mode of action of these compounds, the relative sensitivity of various species or cultivars of a crop species to a given phytotoxin can be determined by using immunofluorescent analysis. Although the results of these experiments are most often described in a nonquantitative manner, counting the cell types in a given configuration (e.g., multipolar mitosis, small tufts of kinetochore microtubules, abnormal spindle arrays) allows for quantitation of the results in terms of cell populations.

COMMENTS

The protocol outlined above is for use with onion root tips or large-seeded grasses. If other plant tissues are to be utilized, variations in specimen preparation prior to labeling, changes in the composition of the solutions utilized in incubation, or subsequent handling of the specimens may result in a better tubulin immunolocalization.

Digestive Enzymes

The protocol of Wick et al. (1981) utilizes a brief incubation in a cellulase preparation to partially digest the onion root tips. Although this protocol is satisfactory for many species, roots of small-seeded grasses and nonroot plant parts of other species generally require a different preparation to maximize the production of free cells. Addition of a pectinase preparation such as Macerase (Calbiochem Corp., LaJolla, California, U.S.), as well as increasing the cellulytic enzyme concentration and time in the digestion step, have been utilized in various other systems with success. A mixed cellulytic digestive agent called Drieselase (Fluka Chemicals, Ronkonkoma, New York, U.S.) has proven very useful in softening cell walls in very difficult to digest systems. Plants with heavy epicuticular waxes may be treated with cutinase or by dipping in organic solvents prior to incubation in the cellulytic enzymes. Many, if not all, of these other preparations are contaminated with proteolytic enzymes, necessitating the addition of BSA or some other protein to the digestion solutions to protect the proteolytically sensitive tubulin. An alternate precaution would be the addition of a cocktail containing a mixture of protease inhibitors.

Detection Systems

Generally, FITC-conjugated secondary antibodies are utilized to identify the sites recognized by the primary antibody because they are readily available from commercial sources and give a very strong signal. Two other choices of fluorescent molecule, rhodamine and Texas red (Worthington Biochemicals, Junction City, Oregon, U.S.), are of use in systems where the endogenous fluorescence is in the same spectral region as FITC. Both of these fluors give a red fluorescence, but are generally less distinct and more difficult to photograph than FITC. They are also disadvantageous in that they fluoresce in the same spectral region as the abundant plant pigment chlorophyll. These alternative fluors have been utilized in dual-labeling experiments in which one protein is labeled with FITC and another labeled with Texas red or rhodamine.

Primary Antibody

Tubulin is a highly conserved protein across phylogenetic lines, so polyclonal sera raised to tubulin from animals and other plant species will generally produce good labeling. Although the original protocol of Wick et al. (1981) utilized polyclonal antiserum, most laboratories now utilize mouse or rat monoclonal antibodies for most immunocytochemical localizations of higher plant tubulin (e.g., Lehnen et al., 1990). The advantages of the monoclonal antisera are consistency and specificity. Because monoclonals are produced from a continuously maintained cell culture, subsequent batches obtained from a given manufacturer will behave very similarly. Polyclonal sera, on the other hand, come from bleeds of a given immunized animal. Different individual animals or subsequent bleeds from the same animal are likely to produce a different population of antibodies. Monoclonal antibodies recognize a single site on the antigen and therefore can be used to identify a single tubulin form (alpha or beta). Polyclonal antisera, which are made of a population of antibodies, recognize both forms because of the significant amino acid homology between the two proteins.

Blocking Agents

Generally, a small amount of BSA is used to block nonspecific antibody binding. It is a relatively inexpensive reagent and is available from a number of sources. Preimmune sera of the animal species used for the secondary antibody or other crude protein homogenates are also effective blocking agents.

Antifading Agents

One of the inherent problems in fluorescence microscopy is that the fluorescence fades irreversibly with continued exposure of the specimen to the excitation light source. Addition of reagents such as propyl gallate to the final mounting medium quenches some of the reactions involved in fluorescence fading and is now standard to immunofluorescence protocols.

Improving the Longevity of Specimens

In the original protocol of Wick et al., the specimens are processed and observed without interruption. Often this is inconvenient or impossible, however. One convenient alternative is to plunge the fully processed slide into liquid nitrogen and then store the slide in plastic boxes at −80°C. The slides are allowed to warm to room temperature before observation. Slides processed in this manner may be observed at least 6 months later with no loss of fluorescent intensity or definition (Sherman and Lehnen, in preparation).

ACKNOWLEDGMENTS

Thanks are extended to Ms. Lynn Libous-Bailey and Ms. Ruth Jones for technical assistance and to Dr. Larry P. Lehnen, Jr. for the modifications of technique used in our laboratory during the course of these experiments. Some of this work was supported by a CRGO grant to KCV.

REFERENCES

Lehnen, L. P., Jr., Vaughn, M. A. and Vaughn, K. C., Terbutol affects spindle microtubule organizing centers. *J. Exp. Bot.,* 41, 537, 1990.

Vaughn, K. C. and Lehnen, L. P., Jr., Mitotic disrupter herbicides. *Weed Sci.,* 39, 450, 1991.

Wick, S. M., Seagull, R. W., Osborn, M., Weber, K. and Gunning, B. E. S., Immunofluorescence microscopy of organized microtubule arrays in structurally stabilized meristematic plant cells. *J. Cell Sci.,* 89, 605, 1981.

CHAPTER 30

Immunoblotting Protocol for the Characterization of Cytoskeletal Proteins Associated with Mitotic Disrupter Herbicides*

Timothy D. Sherman and Kevin C. Vaughn

INTRODUCTION

One quarter of the herbicides presently available affect cell division as a primary mechanism of action (Vaughn and Lehnen, 1991). Herbicidal mechanisms which disrupt mitosis affect microtubule formation and stability. This is affected through herbicidal binding to tubulin or microtubule-associated proteins (MAPs), or through interaction with the spindle or phragmoplast microtubule organizing centers. Other cytoskeletal proteins such as actin, tau, and myosin are potential targets of these herbicides. Once mitotic disruption has been indicated by techniques such as fluorescence microscopy, further biochemical analysis is necessary to ascertain the specific mode of interference and the mechanisms of resistance to these herbicides. Immunoblotting allows for characterization and quantitation of the effected cytoskeletal proteins. Because cytoskeletal proteins are highly conserved between diverse organisms, commercially available monoclonal and polyclonal antibodies to these proteins isolated from other species may be utilized to probe plant extracts. This allows sites of action of mitotic disrupters to be screened rapidly in almost any lab. The procedure outlined here is a compilation of various methodologies. It will produce quality immunoblots of cytoskeletal proteins from a wide range of plant tissues. Modifications to this protocol may improve results in a given system.

PROCEDURE

Equipment

Vertical minigel electrophoresis and electroblotting units (8 × 10 cm format) are ideally suited for this analysis (Hames, 1990). Antibodies which recognize actin, α- and β-tubulin, tau, myosin, or MAPs are available from a number of commercial sources, including Sigma Chemical Company, St. Louis, Missouri, U.S.; Amersham, Arlington Heights, Illinois, U.S.; Polysciences, Inc., Warrington, Pennsylvania, U.S.; Chemicon, Temecula, California, U.S.; and ICN Biomedicals, Irvine, California, U.S.

Tissue Preparation

Plant tissue (root, shoot tips, and etiolated seedlings are good sources of cytoskeletal proteins) are frozen in liquid nitrogen and homogenized with a mortar and pestle. Transfer the tissue, immediately after the liquid nitrogen has totally evaporated, into ice-cold extraction buffer at a ratio of 1 g fresh weight per 5 mL of buffer consisting of 125 m*M* tris(hydroxymethyl)aminomethane (Tris) pH 6.8/10% (v/v) glycerol/2% (w/v) lithium dodecyl sulfate/1% (v/v) β-mercaptoethanol/0.02% (v/v) bromphenol blue/1 m*M* phenylmethylsulfonyl fluoride (PMSF)/10 μ*M* leupeptin/10 μ*M* pepstatin/5 m*M* 1,10-phenanthroline. Heat the sample at 80°C for 10 min. These samples can then be analyzed immediately or stored for longer periods at −20 to −80°C. Avoid storage in frost-free freezers. Repeated freeze/thaw cycles will also render some proteins insoluble and therefore should be avoided.

Electrophoresis

The stock solution of acrylamide is composed of 29.2% (w/v) acrylamide and 0.8% (w/v) bis-acrylamide to give a total acrylamide concentration of 30%. A stacking gel of 4% acrylamide/125 m*M* Tris pH 6.8/0.15% (v/v) *N,N,N'N'*-tetramethylenediamine (TEMED)/0.05% (w/v) ammonium persulfate and a resolving gel of 7.5% acrylamide/375 m*M* Tris pH 8.8/0.05% TEMED/0.05% ammonium persulfate are employed. The electrode buffer consists of 25 m*M* Tris/192 m*M* glycine/0.1% (w/v) sodium dodecyl sulfate (SDS). Electrophoresis is carried out at 100 V until the tracking dye reaches the end of the gel (ca. 2 hr).

Electroblotting and Immunodetection

After electrophoresis, the gel is equilibrated in 100 to 150 mL of 25 m*M* Tris/192 m*M* glycine/0.01% (w/v) SDS/20% (v/v) methanol for 15 to 20 min.

Blotting membrane should be equilibrated similarly in a small amount of buffer. Transfer is at 4°C in the same buffer for 1 hr at 1 mA/cm^2 of gel cassette surface followed by 8 hr to overnight at 3.5 mA/cm^2 (Otter et al., 1987). The finished blot is air-dried to enhance protein binding to the membrane. All subsequent steps can be performed at room temperature. Unoccupied protein binding sites on the membrane are quenched with 0.5% (w/v) Blocking reagent (Boehringer-Mannheim, Indianapolis, Indiana, U.S.) in 10 m*M* Tris-HCl pH 8.0/150 m*M* NaCl (TBS) for 1 hr. Primary antibody (e.g., anti-tubulin) is added at the recommended dilution directly to the blocking solution and allowed to incubate 2 hr to overnight. The blot is then washed three times for 10 min in TBS/0.05% (v/v) Tween-20. Alkaline phosphatase-conjugated secondary antibody complementary to the species of the primary antibody is used to treat the blot for 1 hr. The blot is washed as above. Detection of labeled protein is accomplished with 330 μg/mL nitroblue tetrazolium and 165 μg/ml 5-bromo-4-chloro-3-indolyl phosphate (from 50 mg/mL stock solutions in *N,N*-dimethylformamide) in 100 m*M* Tris, pH 9.5/100 m*M* NaCl/5 m*M* $MgCl_2$. The blot can be developed in this solution for many hours if necessary to enhance weakly reacting bands, as long as it is kept in darkness. Stop color development by washing the blot with running water and then drying it. The blot can be photographed wet or dry. Use of a yellow photographic filter will enhance contrast of the purplish bands and allow for better photography.

RESULTS

Extracts of dodder (*Cuscuta pentagona*) shoots, onion roots, and carrot roots, electrophoresed and blotted as described above, were reacted with monoclonal antibodies to actin, α-tubulin, and β-tubulin (Figure 1). One blot was probed with monoclonal mouse anti-chick gizzard actin, and the other was probed simultaneously with monoclonal mouse anti-chick brain α-tubulin and monoclonal mouse anti-chick brain β-tubulin. The two tubulin antibodes were combined to demonstrate the slight migration difference between the tubulin types, with α-tubulin migrating slightly faster and staining less intensely than the β-form. All of these antibodies recognized single immunoreactive bands in one-dimensional electrophoretic separations. Although these proteins are from different tissues of divergent plant species, they produce bands with very similar relative molecular masses of 43, 53, and 55 kDa for actin, α-tubulin, and β-tubulin, respectively.

Immunoblotting is best if used subsequently to immunofluorescence microscopy (as outlined in Chapter 29). Once disruption of normal cytoskeletal structure is ascertained by immunofluorescence microscopy, characterization of potential anomalies in cytoskeletal proteins should be undertaken. It should be noted that changes in these proteins are not always detected immunologically. Immunoblotting is of value, however, in that it is a rapid method for screening these types of changes, which have been shown to result in herbicide

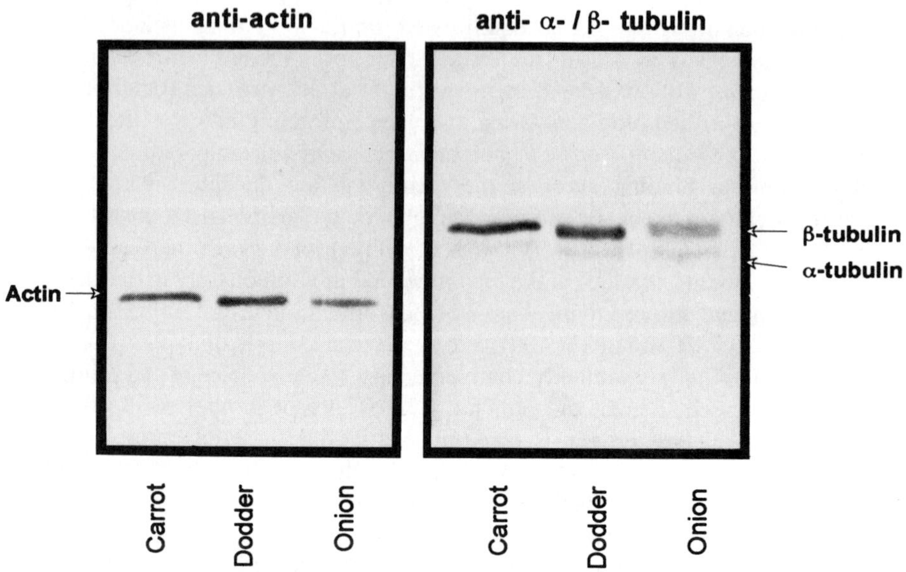

Figure 1. Immunoblot of plant extracts of three different species probed for α-tubulin, β-tubulin, and actin.

resistance in at least some biotypes of weed species (Vaughn and Duke, 1991). Besides the direct effects of microtubule disrupters on microtubule polymerization and location, these compounds also disrupt the production of new tubulin protein. For example, Collis and Weeks (1978) found that amiprophos methyl (APM) disrupted the synthesis of tubulin. This work was conducted prior to the general availability of antisera to tubulin and therefore investigated only effects on tubulin synthesis through the incorporation of radiolabel into new protein, and not stability of tubulin pools. Unfortunately, there have been few other studies that document biochemical effects of mitotic disrupter herbicides.

COMMENTS

Inhibition of proteolytic activity in crude extracts is essential to obtaining useful immunoblots. Samples must be kept frozen at liquid nitrogen temperatures until the addition of protease inhibitors. Concentrated stock solutions (100X) of the protease inhibitors should be prepared in advance and added to the extraction buffer immediately prior to use (PMSF stock solution is prepared in isopropanol, pepstatin and 1,10-phenanthroline in ethanol, and leupeptin in water; pepstatin and leupeptin must be stored at −20°C, while PMSF and 1,10-phenanthroline are stable at room temperature in these solvents.)

For tissues with high concentrations of secondary products, pigments, or salt, or in cases where the protein of interest is in low abundance in the tissue, a slightly modified protocol will give better results. The frozen tissue should be thawed and disrupted in 125 m*M* Tris pH 6.8/1 m*M* PMSF/10 μ*M* leupeptin/10 μ*M* pepstatin/5 m*M* 1,10-phenanthroline. Immediately add four volumes of ice cold acetone, mix, and allow proteins to precipitate at −20°C for 1 to 2 h. Collect the precipitated protein by low-speed centrifugation (3000 × *g* for 5 min). Decant, but do not dry the pellet. Resuspend the moist pellet in the extraction buffer and treat as described above. By decreasing the volume of extraction buffer used in this step, the concentration of the proteins in the sample can be effectively increased by as much as fivefold without degrading the quality of the immunoblot.

The transfer protocol is designed to facilitate the transfer of proteins of a wide range of molecular mass. Smaller proteins bind to the membrane more efficiently at the low current setting. It is also particularly effective in the transfer of myosin (200,000 M_r), which is particularly difficult to remove from the gel matrix. If only moderate to lower molecular mass proteins are to be examined, eliminate the SDS from the transfer buffer and transfer at 1.6 mA/cm^2 for 1 hr.

Nitrocellulose and PVDF membranes give acceptable blots with cytoskeletal proteins. Nylon and charged membranes are less desirable because they are more difficult to block and generally have higher background labeling. Bovine serum albumin [1% (w/v)] or nonfat dry milk [1 to 3% (w/v)] can also be used to block the membrane. The blocking reagent from Boehringer-Mannheim is a nonfat dry milk preparation which has been treated to remove associated lipids and carbohydrates. It yields the lowest background of any blocking reagent utilized in this laboratory. Stocks of these blocking agents containing 0.01% (w/v) sodium azide can be prepared in advance and stored at 4°C.

Horseradish peroxidase- (HRP-) or collodial gold-conjugated secondary antibodies can be used as detection systems in place of alkaline phosphatase- (AP-) conjugated secondary antibody. Although HRP conjugates are less expensive than AP conjugates, they are at least one order of magnitude less sensitive and are inhibited by azide. Collodial gold conjugates, when coupled with commercially available silver intensification kits, are as sensitive as AP conjugates, produce very low background and more easily photographed bands, but are much more costly.

REFERENCES

Collis, P. S. and Weeks, D. P., Selective inhibition of tubulin synthesis by amiprophos methyl during flagellar regeneration in *Chlamydomonas reinhardi*. *Science*, 202, 440, 1978.

Hames, B. D. One-dimensional polyacrylamide gel electrophoresis, in *Gel Electrophoresis of Proteins: A Practical Approach*, Hames, B. D. and Rickwood, D., Eds., IRL Press, Oxford, 1990, p. 1.

Otter, T., King, S. M. and Wiltman, G. B., A two-step procedure for efficient electrotransfer of both high-molecular-weight (> 400,000) and low-molecular-weight (<20,000) proteins. *Anal. Biochem.*, 162, 370, 1987.

Vaughn, K. C. and Duke, S. O., Biochemical basis of herbicide resistance, in *Chemistry of Plant Protection,* Vol. 7, Börner, H., Martin, D., Sjut, V., Stan, H.-J., and Stetter , J., Eds., Springer-Verlag, Berlin, 1991, p. 141.

Vaughn, K. C. and Lehnen, L. P., Jr., Mitotic disrupter herbicides, *Weed Sci.*, 39, 450, 1991.

Model Species and Miscellaneous Topics

CHAPTER 31

Inhibition of Plasma Membrane ATPase

Jean-Pierre Blein and Marie-Louise Milat

INTRODUCTION

The primary active transport process through the plasma membrane involves ATP-dependent electrogenic movements of protons. The electrochemical gradient generated by ATPases provides the driving force for the exchanges between plant cells and the external medium. This gradient can be disturbed by external signals leading to physiological modifications. For example, in plant-microbe interactions, elicitors of hypersensitive reaction induce changes in membrane potential and ions fluxes. It may result from the activation of a plasma membrane K^+ efflux/H^+ influx exchange. Thus, the plasma membrane seems to be involved in signal transmission leading to the induction of defense reaction. In the same way, it has been shown that some herbicides can alter the ionic permeability of the plasmalemma (Blein, 1982) or inhibit its ATPase activity (Watson et al., 1980). Therefore, ATPase appears to be a potential target for phytotoxic molecules.

Plasma membrane can be prepared easily from any material using the phase partitioning method. Then, effects of chemicals on plasma membrane ATPase can be monitored by studying either its phosphohydrolase or its proton transport activity.

Determination of I_{50} values and kinetics analysis of phosphohydrolase inhibition are reported as an example, using two types of molecules: some

0-87371-539-X/93/$0.00 + $.50

carbanilate derivatives (analogous to the herbicide Swep) and beticolin, a toxin from *Cercospora beticola* (for more information see Blein et al., 1987 and 1988).

PROCEDURE

Membrane Preparation

Corn seeds (*Zea mays*) were surface sterilized with calcium hypochlorite (15 g/L, 20 min) and germinated on stainless steel screens above water for 6 days in the dark at 25°C.

All steps were performed at 4°C. Roots (100 g) were homogenized with a Waring blender in 200 mL of a grinding medium containing 50 m*M* Tris/MES 2-[N-Morpholino] ethanesulfonic acid (pH 8.0), 500 m*M* sucrose, 20 m*M* EDTA, 10 m*M* dithiothreitol (DTT), and 1 m*M* phenylmethylsulfonyl fluoride (PMSF). After filtration through Miracloth tissue, the suspension was clarified at 39,000 × *g* for 20 min and the supernatant centrifuged at 140,000 × *g* for 35 min. The 140,000 × *g* pellets were resuspended in 1 mL of 5 m*M* phosphate buffer (pH 6.8) containing 250 m*M* sucrose and 1 m*M* DTT. A 16-g two-phase polymer system was used (1 g of microsomes and 15 g of the polymer mixture), with a final composition of 6.4% (w/w) Dextran T500 and 6.4% (w/w) polyethylene glycol 4000, in the phosphate buffer used above. After mixing the system with 40 inversions, the two phases were separated by centrifugation (1600 × *g*, 5 min). The upper phase was collected and washed with a fresh lower phase. This step was repeated four times, the lower phases being discarded each time. Finally, the upper phase was diluted (five-fold) with resuspension buffer containing 10 m*M* Tris/MES (pH 7.3), 250 m*M* sucrose, 1 m*M* ATP, 2 m*M* EDTA, 1 m*M* DTT, and 1 m*M* PMSF and centrifuged for 35 min at 140,000 × *g*. The membrane pellet was washed once with the same buffer and then resuspended at 1 mg/mL protein in 10 m*M* Tris/MES (pH 7.3), 20% glycerol (w/v), 1 m*M* ATP, 2 m*M* EDTA, 1 m*M* DTT, and 1 m*M* PMSF. The membrane fraction was stored at −80°C. Such preparations can be used up to 1 year (Bourdil et al., 1990).

Phosphohydrolase Activity

Phosphohydrolase activity was measured after a 30-min incubation at 38°C in 1 mL of Tris/MES buffer (pH 6.5) containing 100 m*M* KCl, 3 m*M* $MgSO_4$, 3 m*M* ATP, 5 μg/mL lysolecithin, and 5 to 10 μg membrane protein. The reaction was stopped by adding 3 mL sodium dodecyl sulfate (SDS; 7%, w/v) and the amount of inorganic phosphate liberated was determined. After a 10-min incubation at room temperature with 1 mL of molybdate reagent (5% sodium molybdate in 1*N* H_2SO_4) and 1 mL of an aqueous solution containing 1% Elon and 3% sodium metabisulfite, the sample absorbance at 740 nm was monitored with a colorimeter.

Protein Determination

Protein concentration was determined using bovine serum albumin (BSA) as a standard. Samples (10 μL) diluted with distilled water (1 to 10), were incubated with 2.5 mL Bradford reagent. It contained 100 mg of Coomassie blue (G250) dissolved in 50 mL ethanol; then 100 mL phosphoric acid (85%, w/v) and 850 mL distilled water were added.

Chemicals Used

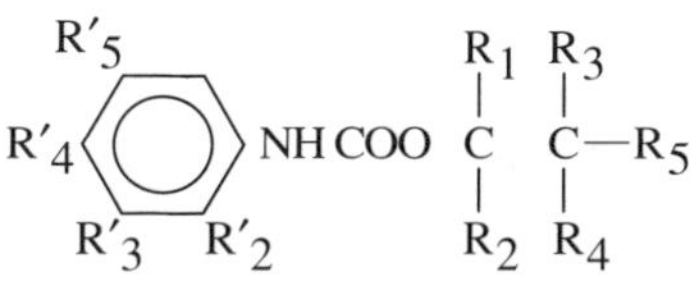

Carbanilate derivatives

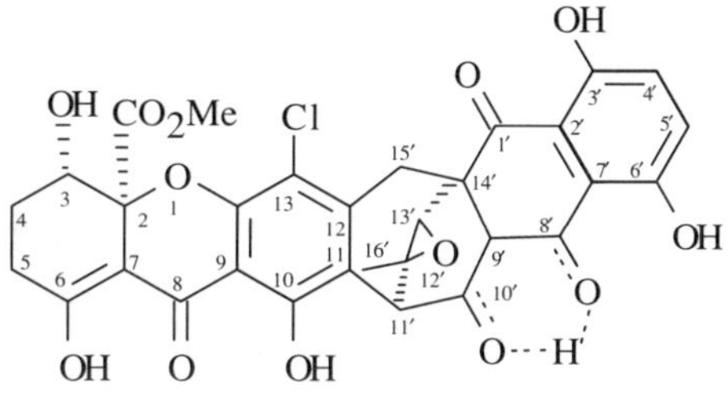

Beticolin 1

RESULTS

Plasma Membrane ATPase

Membranes prepared using the phase partitioning method were at least 95% pure. ATPase activity associated with this fraction was maximum at pH 6.5, sensitive to vanadate or SW26 (2,2,2-trichloroethyl 3,4-dichlorocarbanilate, the first compound listed in Table 1), Mg^{2+}-dependent, and specific for ATP. ATPase activity in a typical experiment was 4.5 μmol ATP hydrolysed per minute per milligram protein, the sample absorbance at 740 nm being about 0.8, using 9-μg protein. From 100 g of roots we usually recovered about 1 mg microsomal proteins.

Table 1. Inhibition of Carbanilate Derivatives on the Phosphohydrolase Activity of the Corn Root ATPase

R′2	R′3	R′4	R′5	R1	R2	R3	R4	R5	%
H	Cl	Cl	H	H	H	Cl	Cl	Cl	70
H	Cl	H	H	H	H	Cl	Cl	Cl	80
H	CH_3	H	H	H	H	Cl	Cl	Cl	35
H	H	H	H	H	H	Cl	Cl	Cl	31
H	Cl	H	Cl	H	H	Cl	Cl	Cl	75
Cl	H	H	Cl	H	H	Cl	Cl	Cl	20
H	Cl	Cl	H	H	H	H	H	H	20
H	Cl	Cl	H	H	H	Br	Br	Br	82
H	Cl	Cl	H	H	H	CH_3	CH_3	CH_3	25
H	Cl	Cl	H	H	CH_3	Cl	Cl	Cl	74
H	Cl	Cl	H	CH_3	CH_3	Cl	Cl	Cl	07

[a] Results are expressed as percent inhibition.

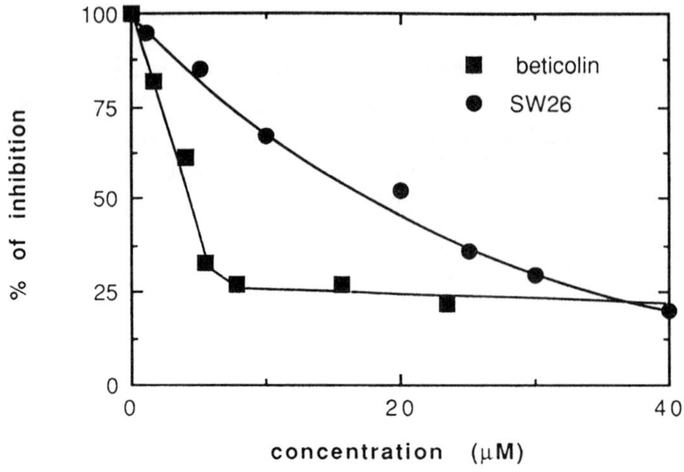

Figure 1. Determination of I_{50} values of phosphohydrolase activity by SW26 and beticolin.

ATPase Inhibitors

When a lot of compounds have to be evaluated, a single concentration can be assayed. In Table 1, the inhibitory activity of 11 carbanilates derivatives (at 100 μM) is reported. These results indicated that a chlorine atom at R′3 and a substitution by a halogen atom at positions R3, R4, and R5 were essential to keep the inhibitory properties mentioned (Blein et al., 1987). These data were qualitative, and it was necessary to determine an inhibition constant.

In order to further study some chemicals, the first approach was to measure the concentration which inhibited 50% of the initial activity (I_{50}). When various concentrations of beticolin or SW26 were added to the membrane fraction, phosphohydrolase activity decreased. Figure 1 shows the relationship between this inhibition and the concentrations used. ATPase appeared to be more sensitive to beticolin (I_{50} = 4 μM) than to SW26 (I_{50} = 20 μM). Although I_{50} values allow a comparison between different molecules, they depend on experimental conditions and cannot be used to characterize a compound. As an example, Figure 2 shows that phosphohydrolase inhibition by beticolin increases with incubation time. Therefore, the inhibition kinetics need to be analyzed in order to determine a more relevant constant.

The most characteristic features of enzymatic kinetics in the presence of varying concentrations of inhibitors are linear double-reciprocal plots 1/V (velocity) vs. 1/S (substrate concentration). The slope, the intercept, or both are altered by a factor $1 + i/K_i$), where i is the concentration of inhibitor and K_i the characteristic constant.

If one considers how an inhibitor might influence the rate of enzymatic reaction involving only one substrate, one can conceive different ways, two of them being illustrated in Figure 3:

1. The straight lines intersect at one point, situated at the left of the origin on the 1/ATP axis. This inhibition is purely noncompetitive, as shown for the inhibition induced by SW26. The inhibitor can combine with the enzyme at a site not identical, wholly or partly, to the substrate-binding site. There is no influence on the affinity of ATPase for ATP.
2. Only the slope is affected by the presence of beticolin. The inhibitor may sufficiently resemble the substrate to be bound in its stead to the active site.

From these plots, K_i values were 15 μM and 6 μM for SW26 and beticolin, respectively.

COMMENTS

Nowadays, ATPase can be prepared more easily from plant material than from cell suspension cultures. Whereas plants can be homogenized in mixers, cells may have to be frozen before being homogenized in a mortar or sonicated. The procedure used to disrupt cells can induce membrane fusions (sonicated cells) and thus lead to plasma membranes contaminated by other endomembranes. The degree of enrichment in plasma membrane ATPase must be controlled for each membrane preparation (see Bourdil et al., 1990).

Since plasma membrane ATPase is a key enzyme for living cells, it is an interesting target for the research of new active compounds. Among them, elicitors of defense reactions seem to be promising tools for phytoprotection.

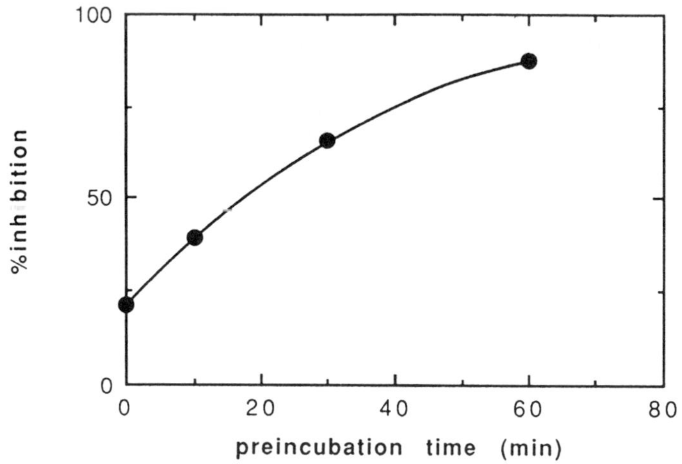

Figure 2. Influence of the preincubation time, in the presence of beticolin, on the I_{50} value.

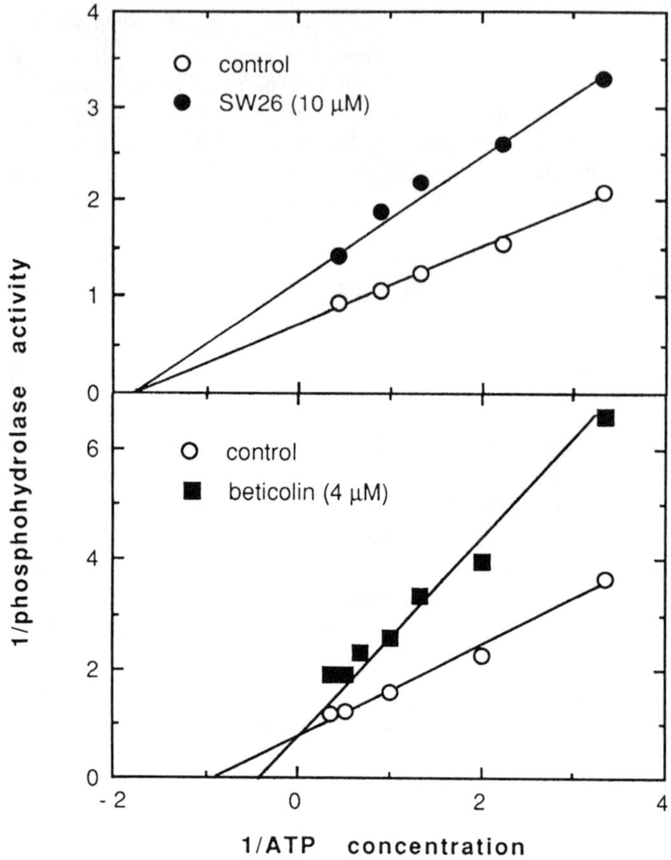

Figure 3. Kinetic analysis of phosphohydrolase inhibition. Phosphohydrolase activity is in micromoles of ATP hydrolyzed per minute per milligram protein, and ATP concentrations are in millimolar units.

REFERENCES

Blein, J.-P., Action of some herbicides on growth, respiration, plasmalemma integrity, and proton extrusion of *Acer pseudoplatanus* cells. II. Amides, diphenyl ethers, nitriles, phenols, triazines, and uracils. *Pestic. Biochem. Physiol.*, 17, 156, 1982.

Watson, M. C., Bartels, P. G. and Hamilton, K. C., Action of selected herbicides and Tween 20 on oat (*Avena sativa*). *Weed Sci.*, 28, 122, 1980.

Bourdil, I., Milat, M.-L. and Blein, J.-P., Effect of lysophosphatidylcholine on the NADH-ferricyanide reductase activity associated with the plasma membranes of corn roots. *Plant Sci.*, 70, 143, 1990.

Blein, J.-P., Martinez, J., Bergon, M., Calmon, J.-P. and Scalla, R., Inhibition of adenosine triphosphatase activity from a plasma membrane fraction of *Acer pseudoplatanus* cells by carbanilate derivatives. *Plant Physiol.*, 83, 469, 1987.

Blein, J.-P., Bourdil, I., Rossignol, M. and Scalla, R., *Cercospora beticola* toxin inhibits vanadate-sensitive H^+ transport in corn root membrane vesicles. *Plant Physiol.*, 88, 429, 1988.

CHAPTER 32

Assays of Xenobiotics on Respiratory Activities in Isolated Plant Mitochondria

Patrick Ravanel and Michel Tissut

INTRODUCTION

Isolated plant mitochondria are the primary site of action of several well-known herbicide families, such as simple phenols (dinoseb, dinoterb, dinitrocresol). Two major effects can be easily detected: an inhibition of the electron transfer and an uncoupling of oxidative phosphorylation. Some compounds give both effects. If a new series of experimental herbicides possesses a potent uncoupling activity in plant mitochondria, further biological investigations will probably be stopped because this effect is likely to induce acute toxicity for animal and human mitochondria. However, looking for new herbicides acting specifically on plant mitochondria remains an attractive aim, as it suggested by the bewildering role played by the toxin of *Helminthosporium maydis,* and some analogues of methomyl, on mitochondria isolated from male sterile (T cytoplasm) maize (Gauvrit and Aranda, 1983).

PROCEDURES

Plant Material

Freshly harvested potato tubers (*Solanum tuberosum* L.) are not appropriate for extractions of tightly coupled mitochondria. Potatoes are obtained from

0-87371-539-X/93/$0.00 + $.50

a local market in October and stocked, after an antisprouting treatment (chlorpropham), in the dark at room temperature. From this material, mitochondria with high respiratory process rates are obtained from November up to the end of January. The end of the dormancy period (February) is not favorable, respiratory controls being very low. From sprouting tubers, organelles of good quality can be isolated, but only during a short period in March or April.

Isolation and Purification of Mitochondria

Fresh material (1 kg), stocked overnight at 4°C, is peeled and cut in small cubes, quickly rinsed in fresh water, and immersed in 1.5 L of chilled medium (4°C) containing 0.3 *M* mannitol, 4 m*M* cysteine, 1 m*M* EDTA, 30 m*M* 4-morpholino-propanesulfonic acid (MOPS) buffer (pH 7.5), and 0.1% defatted bovine serum albumin (BSA). The tuber pieces are mashed with a Moulinex hand grinder for 30 sec. The brei is rapidly seived through 50-μm nylon netting. After an initial low-speed centrifugation (1500 × *g*, 15 min), the mitochondria are sedimented at 10,000 × *g* (20 min). The pellet is suspended in 30 mL fresh medium devoid of cysteine. After a second low-speed centrifugation (1500 × *g*, 15 min) in order to pellet remaining starch grains, mitochondria are resedimented at 10,000 × *g*, giving a pellet of "washed mitochondria" which is kept on ice. This method, when carried out with care, produces mitochondria with a high degree of intactness. The quality of the organelles is largely dependent on experimental details: grinding time, grinding procedure, and pellet dispersion. A further purification procedure, involving centrifugation on sucrose or Percoll gradients, can be used (Douce, 1985).

Equipment

For the extraction procedure a refrigerated centrifuge with a large capacity is needed. The mitochondrial assays require a Clark-type oxygen electrode system and a double-beam spectrophotometer.

Herbicide Solutions, Electrode Medium, and Protein Content

Xenobiotics must be dissolved in dimethyl sulfoxide (DMSO) or ethanol. The respiratory activity is not affected by these solvents for final concentrations below 3% in the reaction medium. O_2 exchanges are followed up at 25°C. The reaction medium (pH 7.2), containing 0.3 *M* mannitol, 5 m*M* $MgCl_2$, 10 m*M* KCl, 10 m*M* Pi buffer and in many cases 0.1% BSA, is carefully oxygenated before the experiment. The brownish pellet of washed mitochondria is suspended with a paintbrush in 300 to 500 μL of fresh extraction medium devoid of cysteine. Protein content is determined spectrophotometrically using Bradford's (1976) procedure. About 0.3 to 1 mg mitochondrial protein is needed per assay (10 to 50 μL of the mitochondrial pellet in 1 mL of electrode medium).

Respiratory Substrates and Inhibitors Needed for Assays

Required substrates and inhibitors include 1 m*M* exogenous NADH; 6 m*M* succinate + 0.3 m*M* ATP; 10 m*M* citrate + 1 m*M* NAD; 10 m*M* α-keto-glutarate + 150 μ*M* thiamine pyrophosphate + 2 m*M* malonate + 1 m*M* NAD; 15 m*M* malate + 1 m*M* NAD; 50 μ*M* ADP for a respiratory control measurement, 2 m*M* for a stable state III; 5 μ*M* CCCP, 5 μ*M* antimycin A, 30 μ*M* KCN, and 10 μ*M* carboxyatractyloside (final concentrations in the reaction medium).

Test Procedures

Three major activities can be detected by polarography:

1. Uncoupling effect on oxidative phosphorylation
2. Inhibitory activity on the system: penetration of substrates + Krebs cycle oxidations + electron transfer
3. Detergent action

Inhibition of Electron Transfer

The best way in which inhibitory activity at the electron transfer level can be detected is to uncouple organelles in the reaction vessel at the beginning of the experiment. In the presence of a reference uncoupler (*m*-Cl phenylhydrazone = CCCP), the oxygen consumption has a maximum rate for energized mitochondria (plus substrate), and it differs according to the oxidized substrate (exogenous NADH: 200 to 250 nmol O_2 per minute per milligram protein; succinate: 200 to 250 nmol O_2 per minute per milligram protein; malate oxidized at pH 6.5: 90 to 120 nmol O_2 per minute per milligram protein; malate oxidized at pH 7.5: 50 to 60 nmol O_2 per minute per milligram protein; α-ketoglutarate: 120 to 150 nmol O_2 per minute per milligram protein) An inhibitory effect taking place between the quinone pool and the end of the electron chain decreases the oxygen consumption rate to the same extent whatever the substrate, as is the case with antimycin A or KCN (inhibitors of complexes III and IV, respectively; see Figure 1). On the other hand, selective or preferential inhibition can be obtained if a xenobiotic acts before or at the level of the flavoproteins. For example, a selective inhibitor of complex II only affects the oxidation of succinate. The oxidation of exogenous NADH can be affected selectively through a specific inhibition of the external NADH dehydrogenase of the inner membrane. This latter oxidation pathway is found only in plant mitochondria. The selective inhibition of mitochondrial complex I is characterized by several tests:

1. No effect can be observed on exogenous NADH or succinate oxidation.
2. Malate oxidation is inhibited at pH 7.5.

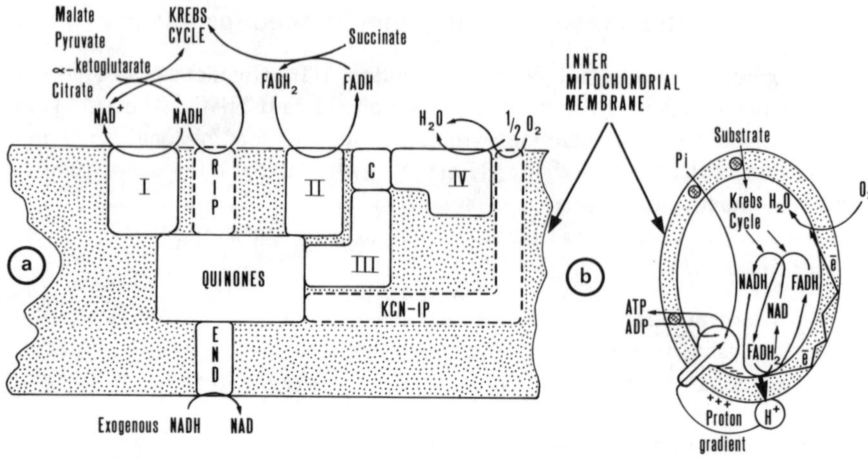

Figure 1. The respiratory chain with its different components in the inner mitochondrial membrane: (a) the electron transfer chain and (b) the oxidative phosphorylation mechanism. I, II, III, IV = complexes; END = exogenous NADH dehydrogenase; RIP = rotenone-insensitive pathway; KCN-IP = cyanide-insensitive pathway; C = cytochrome *c*.

3. There is a one-third decrease of the ADP/O ratio during malate oxidation at pH 6.5 or during α-ketoglutarate, citrate, or pyruvate oxidation at a pH of about 7.
4. During malate oxidation at pH 6.5, a transient inhibition appears. This situation is due to the presence of an alternative rotenone-insensitive pathway in plant mitochondria (for a theoretical explanation see Douce, 1985).

Uncoupling Effect

This test must be carried out using a substrate, the oxidation of which is inhibited poorly or not at all by the molecule tested. After energization of mitochondria and a short state III-state IV transition (30 to 50 μM ADP), carboxyatractyloside can be added in order to inhibit the nucleotide carrier (stable state IV situation). The addition of an uncoupler at this stage disrupts the H^+ gradient, leading to a maximum increase in the O_2 consumption rate (250 to 300% increase). A full uncoupling effect is obtained when the rate of O_2 consumption is no longer stimulated by the addition of CCCP. The uncoupled oxidation rates are almost completely cyanide sensitive in potato mitochondria. These rates are often greater than those of state III. In experiments without carboxyatractyloside, after a full uncoupling effect, a further addition of ADP remains without effect. An uncoupling effect also can be detected by spectrophotometric measurements of the swelling of mitochondria in NH_4Cl or NH_4NO_3 iso-osmotic medium (Ravanel et al., 1986).

Inhibitory Effect on the Phosphorylation Process

If the inhibitions obtained for state III are greater than those obtained for an uncoupled state, the phosphorylation process is probably affected by the molecule under study (Pi or nucleotide transport or ATPase activity). Those inhibitions which specifically affect the oxygen consumption for state III cannot decrease this rate to a lower value than state IV.

Detergent Effect

A detergent effect can be measured by polarography and/or spectrophotometry, using integrity tests for mitochondria. The first assay is based on the reduction of added soluble cytochrome *c* (30 μ*M*) by the mitochondrial cytochrome *c*, in the presence of KCN, the electron donor being succinate (plus ATP). The outer mitochondrial membrane presents an impenetrable barrier to the added soluble cytochrome *c*. An increase in O_2 consumption is correlated with an increase in outer membrane damage. A second assay can be performed with ascorbate (8 m*M*) as electron donor. In the absence of KCN a slow rate of oxidation occurs with intact mitochondria, whereas in the presence of a detergent, the oxygen in the reaction medium is quickly consumed. This oxygen consumption is cyanide sensitive. The third test is spectrophotometric and is based on the reduction of ferricyanide, shown by the decrease (at 418 nm) of the oxidized form peak. About 0.6 mg mitochondrial protein is suspended in 2 mL electrode medium without BSA. The electron donor is succinate (plus ATP) supplemented with antimycin A. In intact mitochondria inhibited by antimycin A, $FeCN_6^-$ is not reduced. If membranes are damaged by detergent xenobiotics, $FeCN_6^-$ is reduced by complex II.

RESULTS

Figure 2 corresponds with the results described below.

Selective Effect of Deguelin on Complex I

Deguelin is a natural rotenoidic polyphenol extracted from roots of *Derris elliptica,* a South American leguminosae. It is a potent inhibitor of complex I in mitochondria (Ravanel et al., 1984). The powder of these roots is used as a biological insecticide due to its toxicity on animal mitochondria devoid of an alternative rotenone-insensitive pathway.

Preferential Inhibitory Effect on the External NADH Dehydrogenase of the Inner Membrane NDE

This situation was demonstrated with many natural and synthetic compounds in this laboratory (including many lipophilic OH-flavones, several

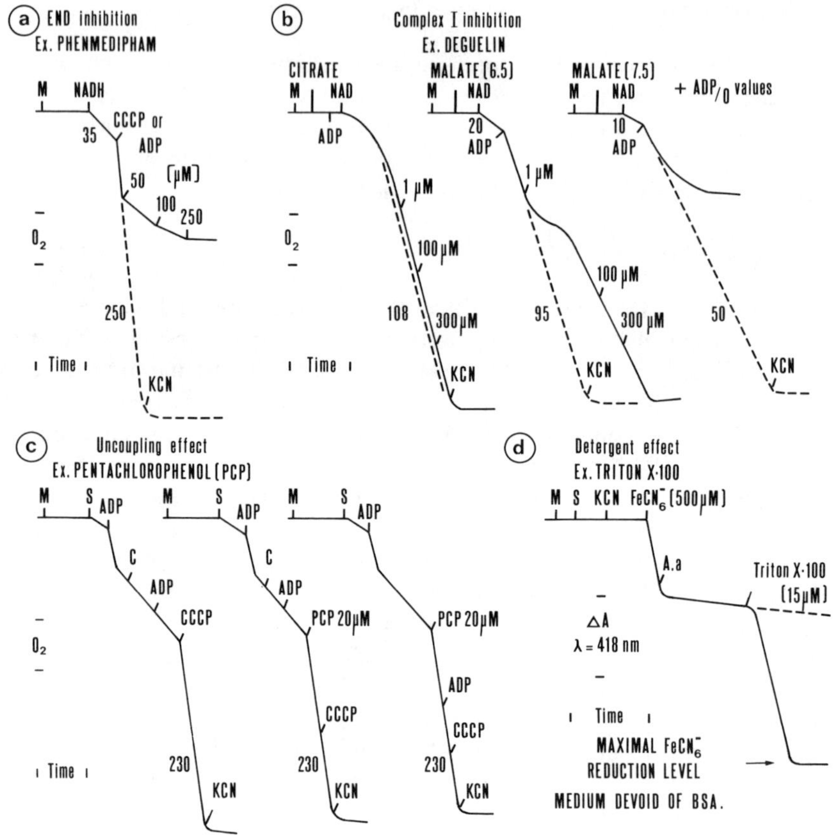

Figure 2. Polarographic and spectrophotometric traces showing: (a) an END inhibition, (b) a complex I inhibition, (c) an uncoupling effect, and (d) a detergent activity on isolated plant mitochondria. M = mitochondria (0.3 to 0.6 mg protein); KCN, 30 μM; S = 6 mM succinate + 0.3 mM ATP; C = carboxyatractyloside (10 μM); A.a. = antimycine A (5 μM); CCCP = m-Cl phenylhydrazone (5 μM). Dotted lines = references traces. Numbers on traces refer to nanomoles O_2 consumed per minute per milligram protein.

phenmedipham derivatives, isoproturon, and some phenylureas and symmetrical diphenylureas). All these compounds affected NADH dehydrogenase first, but for higher concentrations were also able to affect complex I and, to a lesser extent, complex II (Routaboul et al., 1991).

Uncoupling Effect of Pentachlorophenol

The main toxic role of pentachlorophenol on living beings is its uncoupling effect on oxidative phosphorylation (Ravanel and Tissut, 1986). This effect is frequently encountered in the phenol series.

Detergent Effect

This effect was generally observed with all the nonionic detergents, such as Triton X-100 (which had its full effect at 15 μM).

COMMENTS

The comparison of the tests with several Krebs cycle substrates also gives information about substrate transport inside mitochondria and possible inhibition of the carriers and/or the Krebs enzymes concerned. This kind of inhibition was never encountered in our studies.

Some plant mitochondria possess an insensitive KCN pathway branched on the quinone pool. Potato tuber mitochondria are not appropriate for this test. Mitochondria isolated from mung bean (*Phaseolus aureus* L.) etiolated seedlings are preferred. Depending on the substrate oxidized, the rate of this electron pathway may represent 20 to 40% of the state III oxidation rate of a substrate.

Compounds acting inside the mitochondrial inner membrane need to be sufficiently lipophilic in order to be accumulated inside this membrane. This is especially true for potent uncouplers.

These studies have demonstrated that all of the roughly 50 potent oxidative phosphorylation uncouplers studied in this laboratory were also similarly efficient uncouplers of the photophosphorylation activity in isolated thylakoids.

REFERENCES

Bradford, M. M., A rapid and sensitive method for the quantification of microgram quantities of protein utilizing the principle of protein-dye binding. *Anal. Biochem.*, 72, 248, 1976.

Douce, R., *Mitochondria in Higher Plants, Structure, Function and Biogenesis*, American Society of Plant Physiologists Monograph Series, Academic Press, Orlando, FL, 1985, pp. 23, 177.

Gauvrit, G. and Aranda, G., Action of methomyl analogues on maize mitochondria. *Phytochemistry*, 22, 33, 1983.

Ravanel, P., Tissut, M. and Douce, R., Platanetin: a potent natural uncoupler and inhibitor of the exogenous NADH dehydrogenase in intact plant mitochondria. *Plant Physiol.*, 80, 500, 1986.

Ravanel, P., Tissut, M. and Douce, R., Effect of rotenoids on isolated plant mitochondria. *Plant Physiol.*, 75, 414, 1984.

Ravanel, P. and Tissut, M., Toxicity of pentachlorophenol on isolated plant mitochondria. *Phytochemistry*, 25, 577, 1986.

Routaboul, J. M., Mougin, C., Ravanel, P., Tissut, M., Mrlina, G. and Calmon, J. P., Effects of *N,N'*-bis-(4-trifluoromethylphenyl)-urea on isolated plant mitochondria and thylakoid membranes. *Phytochemistry*, 30, 733, 1991.

CHAPTER 33

Rooting and Algal Tests for Herbicide Mode of Action Studies

Carl Fedtke

INTRODUCTION

Many herbicidal targets are now known at the molecular level; i.e., the primarily inhibited enzyme site is known. Some herbicides are not so well understood, and the sensitive enzyme system is unknown in these cases. There is also a need to study and understand the sequence of events which connects the original inhibition with eventual herbicidal actions, which are sometimes growth regulatory or stimulating. Photosystem II-inhibiting herbicides, for example, eventually kill weeds via singlet oxygen, but may at sublethal concentrations induce shade adaptation reactions. Auxin herbicides induce a cascade of physiological reactions which lead to uncontrolled plant growth and finally to plant death. The mechanisms by which acetolactate synthase inhibitors stop plant growth are not known in detail. Sensitive plant systems are therefore required to study (1) unknown modes of herbicide action and (2) the physiology and biochemistry in plants suffering from herbicide action. The test systems which are described here also may be employed to search for herbicidal and/or growth-inhibitory effects of new synthetic or natural compounds. Because of the many different possible herbicidal sites in plant growth and metabolism and in cell and tissue regulation, several test systems are required. *Chlamydomonas reinhardtii* algal cells serve as a simple plant

0-87371-539-X/93/$0.00+$.50

cell model without any tissue relations. A dicot and a monocot plant each are required (soybean and oat), and light vs. dark incubation help in the detection and classification of herbicidal actions. Root regeneration includes nearly all basic regulatory and metabolic steps in cell and tissue growth. Shoot and leaf development give additional visual hints to the mode of action of a candidate compound. Sensitive responses can be studied further in these test systems at the physiological and biochemical level for a deeper understanding of a particular herbicidal mechanism. The tests described here have already been introduced earlier for mode of action studies of graminicides (Fedtke, 1987).

PROCEDURES

Growth of *Chlamydomonas reinhardtii* Algal Cultures (CMY Test)

The algal culture (strain 11-32a from "Algensammlung Göttingen") is grown photoheterotrophically in a Tris-acetate-phosphate medium. The following stock cultures are prepared:

1. Trace element solution (dissolve 22.0 g $ZnSO_4 \times 7\,H_2O$, 11.4 g H_3BO_3, 5.06 g $MnCl_2 \times 4\,H_2O$, 4.99 g $FeSO_4 \times 7\,H_2O$, 1.16 g $CoCl_2 \times 6\,H_2O$, 1.57 g $CuSO_4 \times 5\,H_2O$, and 1.10 g $(NH_4)Mo_7O_{24} \times 4\,H_2O$ in 550 mL H_2O at 70°C; add 50 g EDTA $\times\,Na_2$ dissolved in 250 mL H_2O at 70°C; bring the pH to 6.7 with 20% KOH; bring volume up to 1 L with H_2O; cool down to 20°C; filter; store at room temperature.
2. Phosphate buffer (combine 30 g KH_2PO_4 and 43.4 g K_2HPO_4, each dissolved in 250 mL H_2O, to give pH 7.2)
3. Beijerinck solution (8 g NH_4Cl, 1 g $CaCl_2 \times 2\,H_2O$, and 2 g $MgSO_4 \times 7\,H_2O$ in 1 L H_2O)

For the culture medium, 50 mL Beijerinck solution, 1 mL phosphate buffer, 1 mL trace element solution, and 2.42 g tris (hydroxymethyl)-aminomethane (Tris) are stirred into 800 mL H_2O. The pH is adjusted to 7.2 with glacial acetic acid (ca. 1 mL), and the volume is finally brought to 1 L with H_2O. For algal stock cultures, 5 g Bacto agar are added to 200 mL culture medium, and the mixture is stirred at 90°C until clear (approximately 30 min). The volume is brought to 250 mL with culture medium, and 5-mL portions are transferred into glass vials. The liquid as well as the agar culture medium are autoclaved for 20 min. The glass vials with the agar are allowed to solidify at an angle of 70° to obtain a larger surface for inoculation.

The algal stock cultures on agar medium are grown at room temperature at a north window and are newly inoculated every 4 to 6 weeks. The liquid cultures are grown on a rotary shaker at 100 rpm in 300-mL Erlenmeyer flasks (wide mouth) filled with 150 mL medium. Conditions for synchronous growth are 12 hr of light (8000 lux, Osram HQI bulbs) at 24°C and 12 hr of dark at 18°C. The density of the liquid cultures should not be allowed to exceed 700,000 cells per milliliter in order to keep them in logarithmic growth. The multiplication factor is 4 to 5 in 24 hr, and the inoculation density is generally around 5000 cells per milliliter. For the testing of inhibitory or herbicidal effects, 40 μL algal culture from end log phase as described above are added to 2 mL culture medium and 20 μL herbicidal solution in methanol in 16 × 95 mm glass vials. After 4 days of growth under the conditions described above, but without shaking, the amount of algal growth in the test vials is estimated by chlorophyll analysis. n-Propanol and petrol (1 mL each) are added, and the vials are shaken for 5 min. After centrifugation, chlorophyll is measured in the upper petrol phase at 450 nm by the use of a titertec photometer. The percentage of growth inhibition is calculated for different herbicidal concentrations by comparison with control vials that contain 20 μL methanol. The titertec plate-reading photometer can be coupled to a computer, allowing automatic calculation of the inhibition percentages and the pI_{50} values (negative log of the molar concentration causing 50% inhibition).

Additional information can be obtained from the *Chlamydomonas* test by Coulter counting and sizing of the algal cells (Fedtke, 1981). Growth and division of the cells can be followed in synchronous cultures, and inhibitions can be seen directly within less than 1 hr. Microscopic examination can reveal effects at the cellular level like, for instance, chlorosis or inhibition of cell division with resulting giant multinuclear cells. Another approach to study mode of action with *Chlamydomonas* cultures is to select for resistant cells and to study the resistance mechanism (Fedtke, 1992). A *Chlamydomonas*-based herbicide screening test had already been suggested earlier by Hess (1980).

Rooting of Soybean Cuttings in the Light (SRL Test)

Soybean (*Glycine max* L. Merr. cv. "Hark") seedlings are grown in soil in 20 × 20 cm trays in a 12 hr light/12 hr dark regime at 10,000 lux and 24°C (light)/18°C (dark). For rooting in the light, the seedlings are cut at ground level when ca. 6 days old (ca. 60 mm high, the first true leaf pair not yet developed) and transferred to 16 × 65 mm vials containing 5 mL H_2O and 50 μL herbicidal solution in methanol. Four cut plants are incubated per treatment, each plant in a separate vial. The cuttings are incubated for 3 days in the herbicidal solutions at the growth conditions described above. The light intensity, however, should not be lower than 20,000 lux for a strong response to many of the herbicidal compounds. In order to achieve good transpiration and herbicide uptake, the humidity should be not more than 60% during the light phase. The herbicidal solutions are then replaced by water, and the incubation is continued for another 4 days. After the total incubation time of 7 days the extent of rooting is estimated as a percentage of the controls (root mass). Anomalous rooting can, for instance, be seen with auxin herbicides. In addition, chlorotic, toxic, and other symptoms on the stems and particularly the leaves are recorded.

Other dicot plants can be used instead of soybeans; particularly, the sunflower has been employed (Fabijan et al., 1981). Our experience with these and other plant species is that they may be easily contaminated with soil fungi and that the rooting response is then quite poor.

Rooting of Soybean Cuttings in the Dark (SRD Test)

Soybeans are germinated and grown as described above. For the rooting test in the dark, the seedlings are cut at ground level when ca. 30 mm high (ca. 4 days old, the cotyledon hook not yet opened). Four cut seedlings are transferred to 11 × 75 mm vials completely filled with water containing the herbicide added in methanol to give a final concentration of 1%. The cotyledons are hung on the edge of the vial with the hypocotyl immersed in the solution. After a 24-hr incubation in the dark at 22°C, the cuttings are transferred to a 10-cm petri dish containing a paper disc and 5 mL H_2O. The petri dishes are kept closed by adhesive tape to avoid opening by the pressure of the growing and curling stems during the following incubation. The preparation of the dark incubation experiment and the transfer to the petri dishes are done under normal daylight conditions. After another 6 days in the dark the extent of rooting is estimated as a percentage of the control cuttings.

Rooting of Oat Cuttings in the Light (ORL Test)

Oats (*Avena sativa* L. cv. 'Flämings Nova') are grown at 18°C in the dark in vermiculite in 20 × 20 cm trays. After 7 days, when the seedlings are 50 to 70 mm tall, and before the primary leaf emerges through the coleoptile, the plants are cut 5 mm below the shoot apex, leaving about 4 mm of the mesocotyl at the base of the cut plants. The cuttings are then additionally cut at the coleoptile end to give a total length of 30 mm. Five cuttings are transferred to 16 × 65 mm vials containing 2 mL H_2O and 20 μL herbicidal solution in methanol. The incubation in 12 hr light/12 hr dark at 24°C (light)/18°C (dark) requires a light intensity of at least 30,000 lux (Osram HQI) for full herbicidal action of thiocarbamate herbicides, for instance. After 4 days, the extent of rooting is estimated in percent with reference to the methanol control treatment. Alternatively, a measurement of root length is possible (see the section on oat rooting in the dark). In addition, leaf emergence (in millimeters) is measured from the upper edge of the cuttings. Since leaf emergence is about 5 mm by cell expansion only, an inhibition of more than 80% is very rarely recorded (control leaf growth is generally 25 to 30 mm). Only rapidly killing herbicides (e.g., protoporphyrinogen oxidase inhibitors in strong light) also eliminate this basic expansion growth by very rapid action. Herbicides having growth-inhibitory action only do not inhibit leaf emergence by more than 80 to 85%, even at high concentrations.

Rooting of Oat Cuttings in the Dark (ORD Test)

The growth of the oat seedlings follows the protocol described above. The cutting of the seedlings is similar to the light treatment, but without trimming at the upper end. For the incubation in the dark, five cut plants are transferred into a 16 × 65 mm vial containing water and herbicidal solution as described above. After incubation for 4 days in the dark at 18°C, each cut plant has regenerated two roots, emerging laterally on opposing sides from the shoot apex. Methanol at 20 μL (1%) slightly inhibits root growth. The root length or root mass may be estimated visually and percentage inhibition obtained by simple visual comparison as a rapid and reliable method. More precise values can be obtained by measuring the lengths of the individual roots (ca. 10 to 15 mm) and calculating averages and standard deviations. In particular, smaller differences in, for instance, antagonism experiments can be thereby validated (Fedtke, 1991). It is also possible to cut the roots and obtain fresh and dry weights with a balance which can weigh 0.01 mg.

Test Compounds

Herbicides,inhibitors, and candidate test compounds are dissolved in methanol at 2 mg/mL and added to the incubation media at 1% of the total volume, giving 20 mg/L as the highest concentration. For lower concentrations, the stock solution is diluted with methanol so that 1% of the total volume is always added to achieve the desired final concentration in the test.

Calculations

The tests always result in a set of data relating percent inhibition to concentration for each compound. The resulting inhibition vs. concentration curve should in principle follow the well-known S shape: a negative curvature at low concentrations and a positive curvature at high concentrations, with a nearly linear portion at approximately 50% inhibition. Using a linear correlation with inhibition data between 25 and 75%, the pI_{50} value can be estimated easily. Care should be taken, however, that data points above and below 50% inhibition are included and that the difference between the highest and lowest inhibition is at least 30%. Alternatively, computer programs using regression analysis with nonlinear (sigmoidal) mathematical functions are available for the fitting of S curves which can make use of all data points available. In our experience the pI_{50} values obtained from nonlinear regressions are similar and not more accurate when compared to those obtained from the linear regression method described above. It must be stressed, however, that the pI_{50} values are not really "constants" in a mathematical sense. They depend very much on the quality of the test material and on the exact test conditions. When these are kept constant, the variation of pI_{50} values is about ±0.2 in one test and ±0.4 in independent tests over the year. The variation can be particularly large in the SRL test, where the herbicide uptake strongly depends on transpiration.

RESULTS

Photosystem II Inhibition

Herbicidal inhibitors of photosystem II should not be active in the dark (Table 1). The low pI_{50} values in the SRD and ORD tests therefore reflect metabolic interactions other than photosynthesis inhibition. Photosynthesis inhibitors typically lead to yellowing and bleaching of the leaves in the SRL test and to inhibition of root growth in the ORL test. The inhibition of the rooting response presumably reflects a lack of photosynthate under these conditions. Depending on the water solubility and mobility of the compound, leaf bleaching may be restricted to the veins, may be evenly spread over the entire leaf surface, or may not occur at all. In contrast to the phytoene desaturase inhibitors (see below), photosystem II inhibitor "bleaching" normally does not produce entirely white tissue. Some compounds apparently do not penetrate into the algal cells and hence, are inactive in this test (compare, e.g., acetolactate (ALS) inhibitors).

Protoporphyrinogen Oxidase Inhibition

Inhibitors of this enzyme (PPO) are typically very phytotoxic in the light and much less phytotoxic in the dark. Green plant cells are killed within a few hours in the light, leading to necrotic tissue and cells in all sensitive tests employed. Soybean stems coalesce in the light, primarily at the upper end of the immersed tissue. Further developement of the leaves is therefore halted. Oat leaves likewise become necrotic. Acifluorfen may be inactive in some tests because of poor cell and tissue penetration and inadequate transport.

Phytoene Desaturase Inhibition/Chlorosis Induction

The molecular herbicidal mechanisms leading to "chlorosis" can vary, including, for example, photosynthesis inhibition (see above). Inhibitors of carotenoid biosynthesis which lead to white tissue in the light are included in this section. They are often only poor inhibitors of the rooting response, but can be discerned by their bleaching action in the SRL test. Particularly in this group of compounds, the difficulty of interpreting *in vivo* results with respect to the possible underlying mechanism(s) is obvious. In addition, the compounds may not penetrate, may not be mobile, or may have different side effects in different tests or tissues, and the inhibited pathway may not be so important in the particular tissue.

Table 1. Inhibition Constants (pI_{50} Values) for Herbicides from Different Mode of Action Groups in the Five Described Tests

Mode of Action Group and Selected Herbicides[a]	pI_{50} Values[b]					
	CMY	SRL	SRD	ORL	OLL	ORD
Photosystem II inhibition						
Diuron	6.4	4.6	0	5.4	4.8	0
Methabenzthiazuron	5.9	5.1	0	4.7	0	0
Atrazin	6.3	5.1	0	5.0	0	0
Metamitron	4.3	4.8	0	4.0	0	0
Metribuzin	6.3	5.5	0	5.5	3.9	0
Lenacil	6.4	5.6	3.8	5.5	3.9	0
Phenmedipham	5.8	4.8	0	4.8	4.5	4.5
Ioxynil	0	5.7	0	4.8	4.7	4.8
Bentazon	0	5.6	0	0	0	0
Pyridate	4.6	5.7	0	4.7	0	4.7
Protoporphyrinogen oxidase inhibition						
Acifluorfen	0	5.9	0	0	4.5	0
Bifenox	7.5	6.7	0	5.9	5.2	4.8
Lactofen	6.3	6.7	0	7.2	6.1	5.4
Oxadiazon	5.9	6.0	0	6.6	6.5	3.7
Phytoene desaturase inhibition/chlorosis induction						
Aminotriazole	0	5.1	4.5	0	0	0
Flurtamone	5.3	0	0	6.0	0	4.7
Diflufenican	8.3	0	0	6.3	0	6.3
Dimethazone	0	4.5	0	4.4	0	0
Fluridone	5.2	5.9	0	7.5	0	0
Pyrazolate	4.0	5.7	0	0	0	0
Electron mediators						
Difenzoquat	0	5.5	5.4	5.3	4.5	5.8
Paraquat	7.0	6.7	6.6	6.0	6.3	5.7
Quinonamid	5.9	4.6	0	0	0	0
Respiratory Uncouplers						
Dinoseb	4.4	5.6	4.6	6.1	5.6	4.7
DNOC	0	4.4	3.7	6.3	4.5	4.4
PCP	5.3	5.0	4.8	5.1	4.9	4.7
Acetolactate synthase inhibitors						
Bensulfuron	5.6	5.3	0	0	0	4.3
Chlorimuron	5.6	4.3	0	5.3	0	5.3
Chlorsulfuron	0	7.5	4.2	4.0	0	0
Sulfometuron-methyl	5.5	7.9	4.3	5.0	0	0
Imazapyr	0	4.6	0	0	0	0
Auxin herbicides						
Chloramben	0	4.3	0	6.0	4.5	5.3
2,4-D	0	6.4	4.7	6.2	4.1	5.5
MCPA	0	6.6	4.7	7.3	4.0	5.6
Quinchlorac	0	4.8	4.5	5.6	3.7	5.6
2,4,5-T	0	6.5	4.4	6.1	4.2	5.6
Chloroacetamides and Oxyacetamides						
Alachlor	5.7	5.0	0	5.6	4.3	5.6
Butachlor	6.9	5.7	0	5.8	5.4	5.6
CDAA	4.9	5.4	0	6.2	5.2	6.2
Mefenacet	5.7	4.6	0	6.5	5.6	7.0
Metolachlor	5.3	5.2	0	6.0	4.9	6.2
Herbicides interfering with microtubules						
Amiprophos-methyl	6.5	6.2	5.9	6.6	5.3	6.7
Chlorpropham	6.0	6.0	5.2	6.5	4.6	6.5
Propham	0	4.4	4.2	5.6	4.6	6.4
Fluchloralin	6.0	6.6	5.8	6.7	5.5	6.3
Trifluralin	5.5	6.1	4.9	6.6	5.7	6.7

Table 1 (continued). Inhibition Constants (pI_{50} Values) for Herbicides from Different Mode of Action Groups in the Five Described Tests

Mode of Action Group and Selected Herbicides[a]	pI_{50} Values[b]					
	CMY	SRL	SRD	ORL	OLL	ORD
Thiocarbamates						
Benthiocarb	4.1	0	0	0	5.4	0
Cycloate	4.6	5.5	0	6.0	4.7	5.0
Diallate	4.5	4.7	0	5.8	5.8	4.5
EPTC	0	0	0	4.5	4.7	4.7
Acetyl-CoA carboxylase inhibitors						
Diclofop methyl	0	0	0	7.0	4.7	7.0
Fenthiaprop-ethyl	4.9	4.6	0	7.7	4.7	7.7
Haloxyfop	5.5	4.7	0	7.7	4.9	7.7
Quizalofop	0	0	0	7.8	4.1	8.8
Sethoxydim	5.2	4.4	0	6.7	5.6	6.8
Others						
Anilophos	0	5.6	4.0	7.8	4.4	6.9
Dichlobenil	0	6.3	5.6	6.6	6.2	6.4
Methyldymron	0	5.3	4.6	5.6	0	6.1
Bromobutide	0	5.8	4.9	5.5	0	4.7
Cinmethylin	0	7.6	6.5	7.4	0	7.6
Dalapon	0	4.4	0	0	0	0
Glufosinate	0	5.6	4.8	7.5	4.0	0
Glyphosate	4.5	6.2	5.7	5.2	0	4.4

[a] For the chemical identities of the common names see Anon. 1 and 2 (1985 and 1991); for chemical names also see references given in the Preface of this book.

[b] pI_{50} = negative log of the molar concentration causing 50% inhibition; CMY = *Chlamydomonas reinhardtii* growth test; SRL and SRD = soybean stem rooting tests in the light and dark, respectively; ORL and ORD = oat meristem rooting test in the light and dark, respectively; OLL = oat leaf growth test in the light; 0 = $pI_{50} < 4.0$.

Electron Mediators

Difenzoquat and paraquat are generators of active oxygen species and are therefore very phytotoxic. They may be activated by photosynthetic, respiratory, and other metabolic systems and are therefore (in principle) active in all tests. Quinonamid is a quinone algicide; its high lipophilicity may be the reason for the low activity in higher plant tissue.

Respiratory Uncouplers

The mode of action of this group respresents a long-known cytotoxic principle with little or no selectivity. The variation in activity should therefore reflect (besides inherent activity differences of the tested compounds) the different rates of uptake, mobility, etc.

Acetolactate Synthase Inhibitors

The inherently very high herbicidal potential of these compounds (ALS inhibitors) cannot be seen in most of the employed tests. One reason may be that a block in branched-chain amino acid biosynthesis is not critical in some of the tissues because of the available reserves (e.g., in the SRD test). Chlorsulfuron, which does not inhibit in the *Chlamydomonas* test, has nevertheless been shown to inhibit the ALS extracted from *Chlamydomonas* to the same extent as ALS from other sources — i.e., with an I_{50} of 20 n*M* (pI_{50} = 7.7).

Auxin Herbicides

These compounds are active only in higher plant tissue. Besides the tabulated inhibitions, a peculiar rooting response in the SRL test may be seen with auxin herbicides: the roots are very numerous and small and often form a row on one side of the stem. Because of their mass they push out the epidermis. Sometimes a curling of the stem occurs, but the exact appearance of the response differs somewhat from compound to compound. The auxin-type response, however, can always be recognized.

Chloroacetamides and Oxyacetamides

These herbicides are inhibitors of cell division and cell growth, but the molecular site of action is not known. Since they are mainly graminicides, only the oat rooting tests are highly sensitive. Phytotoxic symptoms have not been observed in these tests. The sensitivity of the *Chlamydomonas* algae can be employed for detailed mode of action studies.

Herbicides Interfering with Microtubules

The herbicides of this group may be easily discerned by the peculiar tissue swelling effect which can be observed at the oat root tip and meristimatic leaf regions. Cell division is stopped, but cell growth may proceed extensively. The resulting large and rounded cells swell the tissue. In *Chlamydomonas* cultures the large cells can be measured by Coulter Channelyzer analysis (Fedtke, 1981). In the tabulated results a broad action in all tests is obvious for these compounds. A difference to other herbicide groups with broad action (electron mediators and respiratory uncouplers) is the lack of phytotoxicity in the present group.

Thiocarbamates

These graminicides may be discerned by their comparatively good inhibitory activity on the oat leaf growth in the light (OLL). High light intensities are necessary to develop this type of action. Thiocarbamate herbicides are known to inhibit leaf wax synthesis and deposition, which may be the reason for the specific action reported here.

Acetyl-CoA Carboxylase (ACCase) Inhibitors

ACCase-inhibiting herbicides are known as specific graminicides with no side activity on dicotyledonous plants. This specificity of action is also reflected in the results reported in Table 1. It is in particular the oat rooting response which is specifically inhibited at concentrations which positively correlate with the herbicidal action at the intact plant level.

Others

Several additional herbicides are of interest which cannot be included in one or the groups formed above. Some of the herbicidal mechanisms in this group are known: as expected, the cellulose biosynthesis inhibitor diclobenil affects all higher plant systems, but not the algae; glufosinate and glyphosate are well-known inhibitors of specific processes in amino acid biosyntheses, while dalapon has been reported to inhibit lipid biosynthesis. The selection of compounds given here is not complete, of course; its purpose is to demonstrate the usefulness of the described tests for the detection, quantification, and tentative classification of many different types of herbicides and herbicide action.

COMMENTS

Taken together, the results of the five described tests will, in general, provide a preliminary estimation of the herbicidal potential of a test compound and by the activity spectrum also will give some indication for a tentative classification. It is, however, obvious from the results shown in Table 1 that not all compounds of a group behave similarly, for any of several possible reasons mentioned previously. For the study of herbicide group behavior it is therefore necessary to measure the response to several representative compounds. As in any test or set of tests, weaknesses occur: in the described tests some slowly acting herbicides which act only after the depletion of some specific reserves produce only weak effects. This refers to some compounds inducing chlorosis and to ALS inhibitors of the imazapyr type, for example. A second usefulness of the described tests may be seen in the availability of a sensitive cell or tissue material for the study of herbicide mode of action with all the implications: primary target interference, metabolic responses, compositional changes and adaptations, and chain of cause-effect responses leading to growth-regulatory and/or phytotoxic responses.

REFERENCES

Anon. 1, Composite list of herbicides. *Weed Sci.*, 33 (Suppl. 1), 5, 1985.

Anon. 2, Glossary of common names and abbreviations of herbicides. *Weed Res.*, 31, 417, 1991.

Fabijan, D., Yeung, E., Mukherjee, I. and Reid, D. M., Adventitious rooting in hypocotyls of sunflower (*Helianthus annuus*) seedlings. I. Correlative influences and developmental sequence. *Physiol. Plant.*, 53, 578, 1981.

Fedtke, C., Modes of herbicide action as determined with *Chlamydomonas reinhardtii* and Coulter counting, in *Biochemical Responses Induced by Herbicides,* Moreland, D. E., St. John, J. B. and Hess, F. D., Eds., ACS Symp. Ser. No. 181, American Chemical Society, Washington, D.C., 1981, p. 231.

Fedtke, C., Physiological activity spectra of existing graminicides and the new herbicide 2-(2-benzothiazolyl-oxy)-*N*-methyl-*N*-phenylacetamide (mefenacet). *Weed Res.,* 27, 221, 1987.

Fedtke, C., Antagonistic interactions of mefenacet with inhibitors of monooxygenases. *Z. Naturforsch.,* 46c, 915, 1991.

Fedtke, C., Mode of action studies with mefenacet. *Pestic. Sci.,* 33, 421, 1991.

Hess, F. D., A *Chlamydomonas* algal bioassay for detecting growth inhibitor herbicides. *Weed Sci.,* 28, 515, 1980.

CHAPTER 34

Growth Rate Determination of *Lemna* by Video Scan of the Leaf Surface Area

Günter Retzlaff

INTRODUCTION

Several species of the Lemnaceae duckweed are excellent subjects for bioassays to screen for the herbicidal activity of chemical compounds. Compared with more differentiated higher plants, *Lemna* offers several practical advantages. The plants grow rapidly and are easy to manipulate under aseptic conditions. They propagate vegetatively and thus provide genetically homogeneous material. Duckweed grows well in a nutrient solution and the size of the plants allows the use of small petri dishes for trials. Consequently, only small amounts of test substances are required and they are homogeneously distributed to all the plants within one trial. In contrast with other higher plants, the relatively simple anatomy of *Lemna* minimizes the influence of factors such as cuticle penetration and translocation, thus making results more reproducible and easier to interpret. One of the simplest indicators of the influences of chemical substances on the physiological pathways of plants is growth. Growth rates of *Lemna* are normally estimated by counting the number of plantlets in each trial at a sequence of time intervals. This chapter describes a more sensitive, automated system in which growth rates are calculated from the increase in total leaf surface area as determined by a computerized video image analyzer which also calculates the effects of pesticides relative to untreated control cultures.

0-87371-539-X/93/$0.00 + $.50

PROCEDURE

Plant Material

Many species of *Lemna* are appropriate for the test. The experiments reported here were made with an aseptic culture of *Lemna paucicostata* obtained from the collection of Professor R. Kandeler, Universität für Bodenkultur, Vienna, Austria.

Aseptic Culture

Aseptic cultures of various *Lemna* species can be established by washing the plants for 2 to 5 min in an aqueous solution of 1% sodium hypochlorite containing a few drops of the surfactant Tween 80. The plants are then rinsed three times with sterile water and cultivated separately under sterile conditions.

Cultivation

The nutrient solution contained: $MgSO_4 \times 7\ H_2O$, 0.62 g/L; $CaCl_2 \times 2\ H_2O$, 0.54 g/L; KNO_3, 0.4 g/L; KH_2PO_4, 0.2 g/L; $MnCl_2 \times 4\ H_2O$, 0.47 mg/L; H_3BO_3, 0.5 mg/L; $Na_2MoO_4 \times 2\ H_2O$, 0.12 mg/L; $ZnSO_4 \times 7\ H_2O$, 50 μg/L; $CuSO_4 \times 5\ H_2O$, 25 μg/L; $CoCl_2 \times 6\ H_2O$, 25 μg/L. The medium was adjusted to pH 5.5 and autoclaved for 30 min at 110°C. After autoclaving, a 7.3% solution of NaFe-EDTA was added through a sterile filter at a concentration of 100 μL/L sterilized nutrient solution.

Stock cultures of *Lemna* were maintained in 250-mL Erlenmeyer flasks containing 150 mL of the nutrient medium supplemented with 10 g/L sucrose. The plants were grown in an incubator (Rubarth Apparate GmbH, D-3000 Hannover, Germany) at 27 ± 1°C under continuous illumination with daylight-type tubes of approximately 65 μEinstein (μE) $m^{-2}s^{-1}$. At least 5 days before the start of each experiment the plants were transferred to sucrose-free medium.

Chemicals

Samples of unformulated herbicides were obtained from Riedel-de Haën (D-3016 Seelze, Germany) and BASF Aktiengesellschaft (D-6700 Ludwigshafen, Germany). The stock solutions of the chemical compounds were prepared in acetone 100-fold concentrated relative to the final concentration in the experiment.

Test Conditions and Equipment

The tests were carried out in small crystallization dishes (5 cm diameter × 3 cm deep) containing 25 mL of the nutrient solution and a range of concentrations of the test substances added in a maximum of 250 μL acetone.

Two plants from the sucrose-free stock culture were placed in each dish, which was then covered with the lid of sterile petri dish. All experimental cultures were grown for at least 10 days in an incubator under the conditions described above.

The changes in total leaf area in each trial were determined at regular intervals. For these measurements the crystallization dishes were placed on a light table. The leaf area was then recorded by a video camera with close-up optics in combination with a computer containing a frame grabber and appropriate software (System IMAGO, COMPULOG GmbH, D-7030 Böblingen, Germany). The effects of pesticides on growth rates, relative to untreated control cultures, were calculated from the video images.

RESULTS

Table 1 shows the effects of a number of selected herbicides of different chemical classes (Imamura et al., 1991) and biochemical modes of action (Fedtke, 1982; Böger and Sandmann, 1989) on the growth of *L. paucicostata*. The activity of the compounds is expressed as the molar concentration of herbicide required to inhibit plant growth by 50% (I_{50}). The results indicate that most of the compounds induced this degree of growth inhibition in the range between 10^{-6} and 10^{-7} *M*. Alachlor and the urea derivatives chlotoluron and diuron displayed stronger than average biological activity, and the diphenylethers, imidazolinones, and sulfonylureas were the most potent in this test. These compounds had I_{50} values of 10^{-8} to 10^{-9} *M* and below, whereas chloridazon, *S*-ethyl dipropylthiocarbamate (EPTC), dicamba, mefluidide, and the gramincides diclofop methyl and sethoxydim required concentrations above 10^{-5} *M* to induce 50% growth inhibition of *L. paucicostata*.

In addition to its value in purely quantitiative studies, *Lemna* is also useful as a qualitative indicator of the specific physiological effects of new classes of biologically active compounds. For example, the type of damage symptoms caused by a compound, such as bleaching, greening, leaf-size reduction, etc., may provide the first clue to its mode of action at the biochemical level. Diagnosis can be aided further by the fact that, in these tests, *Lemna* is grown in a sterile minimal medium. If the damage caused by a compound can be reversed by supplementing the medium with an essential metabolite, this provides good evidence that the target site of the test compound may be in the biosynthetic pathway of the metabolite. By using a rationally selected range of metabolites from different physiological compartments, it is often possible to eliminate many potential target sites at an early stage and thus focus investigations on the defined area of metabolism most likely to yield useful results.

An example is given in Figure 1, which shows the phytotoxic effects of chlorsulfuron, the active ingredient in ''Glean'', an inhibitor of acetolactate synthase in plants which blocks the biosynthesis of leucine, isoleucine, and

Table 1. Effects of Selected Herbicides on Growth of Duckweeds (*Lemna paucicostata*)

Chemical Class/Common Name[a]	Molar Concentration Required for 50% Growth Inhibition
Carbamate/biscarbamate	
Chlorpropham	3.5×10^{-6}
Phenmedipham	2.2×10^{-7}
Swep	2.4×10^{-6}
Chloracetamide	
Alachlor	3.4×10^{-8}
Metazachlor	3.8×10^{-7}
Cyclohexenone/propionic acid	
Diclofop methyl	1.6×10^{-5}
Sethoxydim	3.1×10^{-5}
Dinitroaniline	
Fluchloralin	1.9×10^{-7}
Pendimethalin	1.4×10^{-7}
Diphenylether	
Na-Acifluorfen	6.0×10^{-8}
Oxyfluorfen	3.8×10^{-9}
Hydroxybenzonitrile	
Bromoxynil	2.4×10^{-7}
Ioxynil	3.8×10^{-7}
Phenoxy acid	
MCPA	8.0×10^{-6}
2,4-D	3.4×10^{-6}
Pyridazinone/pyridazone	
Chloridazon	2.2×10^{-5}
Norflurazon	2.9×10^{-7}
Pyridate	
Sulfonylurea/imidazolinone	
Chlorsulfuron (Glean)	$<10^{-9}$
Imazaquin	2.8×10^{-8}
Thiocarbamate	
Diallate	7.8×10^{-6}
EPTC	6.0×10^{-5}
Triallate	5.2×10^{-6}
Triazine/triazinone	
Atrazine	3.2×10^{-7}
Metamitron	6.5×10^{-6}
Metribuzin	2.1×10^{-7}
Prometryn	1.7×10^{-7}
Urea	
Chlorotoluron	4.0×10^{-8}
Diuron	4.7×10^{-8}
Methabenzthiazuron	2.6×10^{-7}
Quarternary ammonium salts	
Difenzoquat	1.7×10^{-7}
Paraquat	3.4×10^{-8}
Miscellaneous compounds	
Bentazon	4.3×10^{-6}
Dicamba	7.5×10^{-5}
Dichlobenil	3.3×10^{-7}
Glyphosate	2.3×10^{-5}
Glufosinate	2.9×10^{-6}
Mefluidine	1.3×10^{-5}
Lenacil	2.5×10^{-7}
Propanil	2.8×10^{-7}

[a] See Böger and Sandmann (1989) for chemical names.

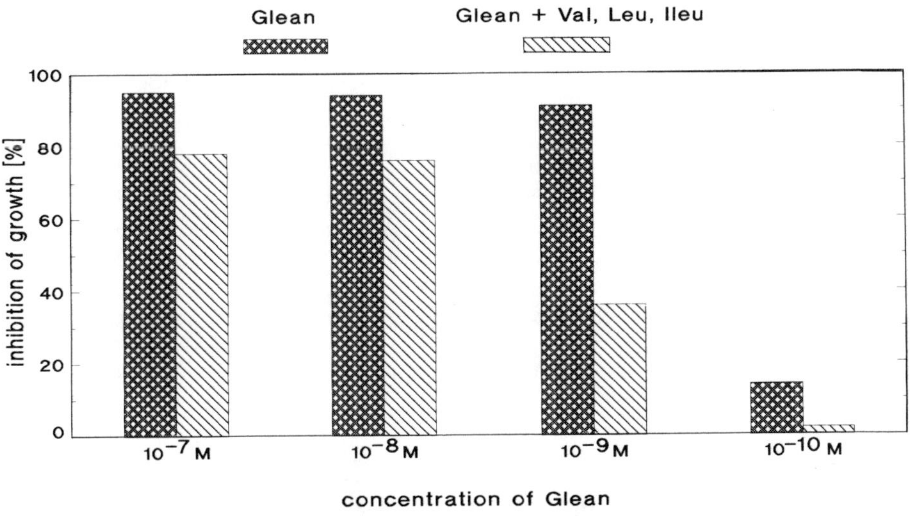

Figure 1. Effect of Glean (active ingredient: chlorsulfuron) on the growth of *Lemna paucicostata* cultivated in the presence and absence of 10^{-4} *M* valine, leucine, and isoleucine in the nutrient medium.

valine. The biological activity of chlorsulfuron is reduced when the plants are provided with a nutrient solution containing these amino acids.

COMMENTS

The advantages of an *in vitro Lemna* test system as a complement to standard greenhouse screening procedures for the determination of phytotoxicity of chemical substances are described. In standard tests new compounds may show no biological activity because they fail to penetrate the plant cuticle or are poorly translocated. The growth of *Lemna* immersed in a liquid medium to which test substances can be directly added, combined with its simple morphology and small size, largely eliminates these problems. The biological acitvities of new toxophores can thus be detected when they would otherwise be overlooked. Other potentially active substances may be overlooked because they are degraded by microorganisms on the leaf surface or in the soil. The sterile medium used in the *Lemna* test system also eliminates this problem.

The small scale of the *Lemna* system requires only minimal amounts of test substances, thus allowing the screening of compounds which are only available in submilligram quantities. The combination small scale and simplicity also produces extremely reproducible and easily interpreted results. Reliable quantitative structure-activity relationship studies which detect subtle differences between compounds of the same chemical class can thus be carried out.

ACKNOWLEDGEMENT

The author thanks Mrs. Sigrid Hennig and Mrs. Anne Keller for technical assistance.

REFERENCES

Imamura, S., Taniyama, H. and Isogawa, T., Eds., *Short Review of Herbicides and PGRs,* Hodogaya Chemical Co., Japan Zenkoku Noson Kyoiku Kyokai Publishing Company, Tokyo, Japan, 1991.

Fedtke, C., Ed., *Biochemistry and Physiology of Herbicide Action,* Springer-Verlag, Berlin, 1982.

Böger, P. and Sandmann, G., Eds., *Target Sites of Herbicide Action,* CRC Press, Boca Raton, FL, 1989.

CHAPTER 35

The Paraquat Bioassay to Evaluate Photosynthesis Inhibition

Daisuke Yanase and Akihide Andoh

INTRODUCTION

Photosynthesis inhibitors are known to antagonize dipyridinium herbicides such as paraquat both in intact plants and under *in vitro* conditions (Yanase et al., 1990; Mees, 1960; Yanase and Andoh, 1989). This phenomenon can be clearly understood to mean that photosynthesis inhibitors reduce the electron supply from the photosynthetic electron transport system to paraquat molecules and consequently decrease destructive superoxide radical production in plant tissues. This principle suggests to us that it is possible to evaluate photosynthetic electron transport inhibition quantitatively by measuring alleviation of paraquat phytotoxicity.

Paraquat kills plants through cellular disruption which can be detected in terms of electrolyte leakage from injured cells measured as the rise in conductivity in the solution bathing plant tissues (Vanstone and Stobbe, 1977). This technique to estimate the phytotoxic injury by paraquat can be integrated into the assay system based on the aforementioned principle. Thus, a new simple but sensitive method to evaluate photosynthesis inhibitors has been established.

This chapter summarizes the standard assay procedure with optimized conditions. Details about establishing this method are described by Yanase et al. (1990).

0-87371-539-X/93/$0.00 + $.50

PROCEDURE

Cucumber Cotyledon Discs

Leaf discs measuring 10 mm in diameter are prepared from fully expanded cotyledons of 8-day-old cucumber seedlings. Ten cotyledon discs are floated on 20 mL of deionized water containing 10 μM paraquat. Compounds to be tested are added to this solution, usually at three or more concentrations. A small amount of acetone (final concentration less than 0.2%) can be used to dissolve compounds having poor water solubility. Each treatment consists of three replicates. These preparations are to be stored in the dark until exposure to light so that paraquat and the test compounds will penetrate the cotyledon discs without injuring them.

Test Compounds

As for test compounds, their chemical or metabolic degradation during the dark preincubation must be taken into consideration. Besides, uncouplers are considered not suitable for this method since in the preliminary experiment pentachlorophenol (PCP) caused electrolyte leakage in the dark preincubation and made the following measurement impossible.

Since alleviation of paraquat phytotoxicity was not detected for some 30 other nonphotosynthesis-inhibiting herbicides tested in preliminary experiments, the present method is highly specific to photosynthesis inhibitors.

In cases where pure compounds were not available, formulated herbicides were employed instead, but so far few problems have been caused by formulating materials.

Measurement of Conductivity Change

A conductivity meter (in this laboratory, Toa, CM-20S) is the only equipment necessary in the present method. The rise in conductivity in the bathing solution is attributable to electrolyte leakage from the cucumber cotyledon discs injured by paraquat phytotoxicity.

The experimental preparations of cucumber cotyledon discs are kept in complete darkness for about 14 hr and then transferred into continuous white light of 81 ± 8 μEinstein (μE) m^{-2} sec^{-1}. The conductivity of test solutions is measured at the beginning of the light exposure and every 2 hr afterward. All these measurements were carried out at 25°C.

The Photosynthesis Inhibition Index, pI_{50} (PQ)

The alleviation of paraquat phytotoxicity by photosynthesis inhibitors is indicated as the decrease of electrolyte leakage for the first 6 hr of light exposure in which the conductivity rises constantly. The photosynthesis-inhibiting activity is expressed in terms of pI_{50} (PQ), defined as the negative

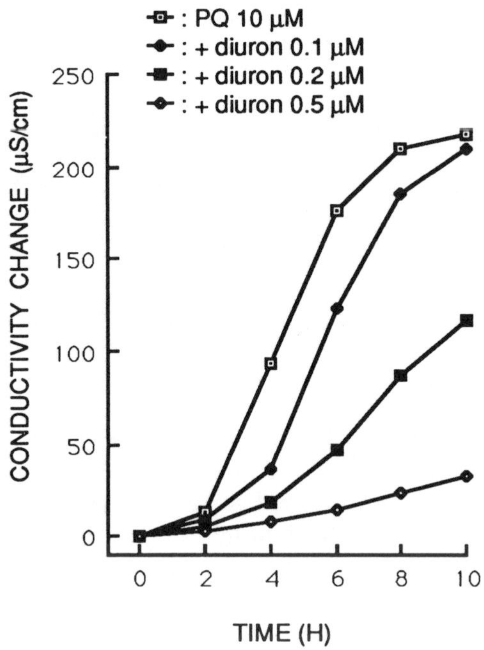

Reduction of paraquat-induced electrolyte leakage for 6 hr (%)

Conc. of diuron (μM)	Trial 1st	2nd	3rd
0.1	- - -	30.2	27.3
0.2	72.6	73.1	64.0
0.5	94.1	91.4	90.4
1.0	95.7	- - -	- - -

$I_{50} = 0.144$ (μM)
$pI_{50} = -\log(I_{50}) =$ **6.84**

Figure 1. Determination of pI_{50} (PQ) for diuron.

logarithm of I_{50} (PQ), which is the molar concentration required to reduce electrolyte leakage by 50% relative to that of the control, in which cotyledon discs are treated only with paraquat. I_{50} (PQ) is determined by probit analysis on the data from assays repeated at least twice for each compound. In this study, the maximum concentration for each of the tested compounds did not exceed 100 μ*M*, so pI_{50} (PQ) values less than 4 could not be measured.

RESULTS

The Conductivity Rise Caused by Paraquat Phytotoxicity

The conductivity of bathing solutions at the beginning of light exposure was usually 30 to 40 μS cm^{-1}. Conductivity increased by about 150 μS cm^{-1} for the first 6 hr of light exposure when cotyledon discs were treated solely with 10 μ*M* paraquat, whereas the rise in conductivity for the same period was about 10 μS cm^{-1} in the dark.

Determination of the pI_{50} (PQ) for Diuron

Figure 1 shows conductivity changes in preparations where cucumber cotyledon discs were treated with 10 μ*M* paraquat alone or together with 0.1, 0.2, or 0.5 μ*M* diuron. Lines in this figure suggest that the I_{50} (PQ) for diuron falls somewhere between 1×10^{-7} and 2×10^{-7} *M*. The calculated values of I_{50} (PQ) and pI_{50} (PQ) for diuron were 1.44×10^{-7} *M* and 6.84, respectively.

pI_{50} (PQ) for Representative Photosynthesis Inhibitors

The pI_{50} (PQ) values for other representative photosynthesis-inhibiting herbicides were presented in Table 1. These values agree well with the pI_{50} values reported by other researchers working with conventional methods. The fact that most of the pI_{50} (PQ) values are greater than 5 suggests that this method is sensitive enough to detect photosynthesis inhibitors at about 10^{-6} *M*.

Evidence from Phytofluorography

It is widely accepted that when the photosynthetic electron transport system encounters interference, plant tissues re-emit photo energy once absorbed by chlorophylls as a form of fluorescence known as chlorophyll fluorescence (Richard et al., 1983). In the experiment described in this chapter, a close relationship was observed between the decrease of electrolyte leakage and the intensity of chlorophyll fluorescence in cotyledon discs treated with various concentrations of a photosynthesis inhibitor (Figure 2). This observation vindicates the relevance of our experimental system.

COMMENTS

This chapter has introduced a rapid and quantitative semi-*in vitro* measurement of photosynthetic electron transport inhibition. Compared with conventional techniques such as the Hill reaction measurement and measurements of photosynthetic oxygen evolution, our method has the following advantages: it requires neither expert skill nor expensive equipment; and, since it consists of synchronous assays on cotyledon discs, this method provides photosynthesis inhibition indices more closely related to actual phytotoxicity of test compounds than values obtained with such materials as isolated chloroplasts or cultured cells.

Some pitfalls of this method have already been suggested in the section on test compounds. Since dark preincubation is necessary so that the test compounds will distribute in cotyledon discs uniformly, decomposition of these compounds either in water or in cotyledon discs must be taken into consideration. However, from the practical point of view, good herbicidal performance cannot be expected from such unstable compounds that would decompose easily overnight. Actual herbicidal activity, then, is expected from compounds with high pI_{50} (PQ) values so long as they can enter the plant apoplast either via roots or via aerial parts.

Although it is a rather primitive assay, the present method is highly sensitive and versatile. In the authors' laboratory, this method is used not only in evaluating herbicide candidates, but also in estimating soil adsorption of these compounds. A future prospect is the expansion of this method to evaluate crop selectivity of photosynthesis-inhibiting herbicide candidates using plant materials other than cucumber.

Table 1. pI_{50} (PQ) Values for Representative Photosynthesis Inhibitors

Herbicide	pI_{50}
Urea derivatives	
Chlorbromuron	5.98
Chlorotoluron	5.34
Chloroxuron	5.45
Cycluron	4.90
Diuron	6.84
Fenuron	4.36
Isoproturon	5.90
Isouron	6.29
Karbutilate	6.40
Linuron	6.72
Methabenzthiazuron	4.98
Metoxuron	5.28
Monolinuron	5.26
Monuron	5.24
Neburon	6.28
Tebuthiuron	6.07
Thiazafluron	5.92
Triazine derivatives	
Ametryn	5.85
Atrazine	6.22
Cyanazine	5.65
Desmetryn	5.51
Dimethametryn	5.80
Dipropetryn	6.00
Methoprotryne	5.98
Prometon	5.29
Prometryn	5.42
Simazine	5.19
Simetryn	6.27
Terbumeton	5.64
Terbuthylazine	5.71
Terbutryn	6.06
Others	
Bentazon	5.21
Bromacil	6.48
Busoxinone	6.45
Buthidazole	6.46
Hexazinone	6.54
Isomethiozin	5.24
Lenacil	6.18
Metamitron	4.99
Metribuzin	6.59

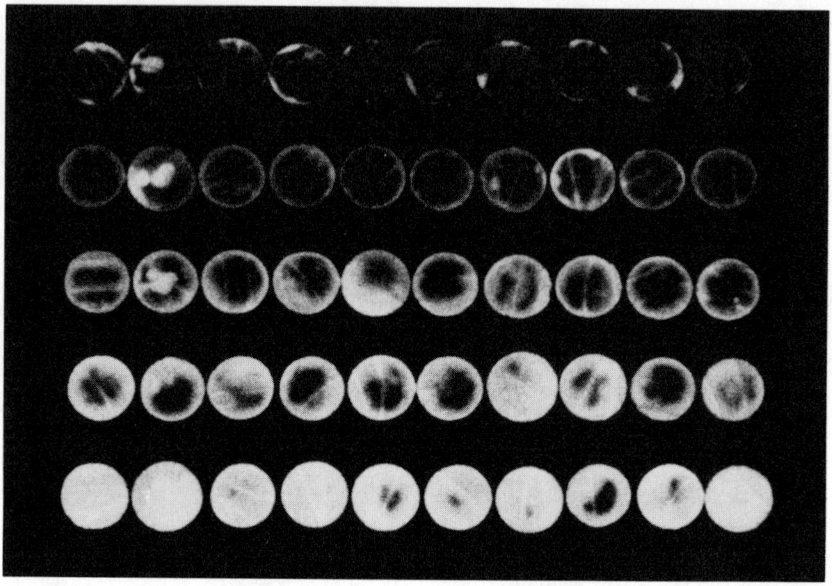

Figure 2. The chlorophyll fluorescence in cucumber cotyledon discs treated with 0, 0.2, 0.5, 1.0, and 2.0 μM isouron (top to bottom), where the photosynthetic electron transport was reduced by 0, 18, 43, 69, and 84%, respectively, according to the paraquat bioassay. A detailed report on our chlorophyll fluorescence studies is now in press. (From Yanase, D. and Andoh, A., *Pestic. Biochem. Physiol.,* in press. With permission.)

REFERENCES

Yanase, D., Andoh, A. and Yasudomi, N., A new simple bioassay to evaluate photosynthetic electron-transport inhibition utilizing paraquat phytotoxicity. *Pestic. Biochem. Physiol.,* 38, 92, 1990.

Mees, G. C., Experiments on the herbicidal action of 1,1′-ethylene-2,2′-dipyridylium dibromide, *Ann. Appl. Biol.,* 48, 601, 1960.

Yanase, D. and Andoh, A., Porphyrin synthesis involvement in diphenyl ether-like mode of action of TNPP-ethyl, a novel phenylpyrazole herbicide. *Pestic. Biochem. Physiol.,* 35, 70, 1989.

Vanstone, D. E. and Stobbe, E. H., Electrolytic conductivity — a rapid measure of herbicide injury. *Weed Sci.,* 25, 352, 1977.

Richard, E. P., Gross, J. R., Arntzen, C. J. and Slife, F. W., Determination of herbicide inhibition of photosynthetic electron transport by fluorescence. *Weed Sci.,* 31, 361, 1983.

CHAPTER 36

The Starch Assay in Leaf Segments

Takao Kida and Hiroshiro Shibai

INTRODUCTION

Photosynthesis is one of the most important physiological targets of herbicides because of the absolute selective toxicity that this produces between plants and animals. Photosynthesis inhibitors prevent oxygen evolution, electron transport, and carbon dioxide fixation, and these effects inhibit the synthesis of starch. Therefore, it is expected that photosynthesis inhibitors are among the compounds which inhibit *de novo* starch synthesis. Starch is accumulated in leaves through photosynthesis in some kinds of plants and can be detected by iodine staining after removal of chlorophylls with hot methanol. This chapter describes attempts to devise a new assay method for detection of photosynthesis inhibitors by determining *de novo* starch synthesis in excised leaf segments.

PROCEDURE

Plant Species

Italian ryegrass (*Lolium multiflorum* LAM.) and barnyard millet (*Panicum crus-galli*) were used as C_3 and C_4 plants, respectively.

0-87371-539-X/93/$0.00 + $.50

Herbicides

The different kinds of herbicides used were diuron (3-(3,4-dichlorophenyl)-1,1-dimethylurea), propanil (3′,4′-dichloropropionanilide), Swep (methyl *N*-(3,4-dichlorophenyl)carbamate), linuron (3-(3,4-dichlorophenyl)-1-methoxy-1-methylurea), paraquat (1,1′-dimethyl-4,4′-bipyridinium dichloride), glyphosate (isopropylamine salt of *N*-(phosphonomethyl) glycine), nitrofen (2,4-dichloro-1-(4-nitrophenoxy)benzene), 2,4-D (2,4-dichlorophenoxyacetic acid), amitrole (3-amino-1H-1,2,4-triazole), chlorpropham (isopropyl *N*-(3-chlorophenyl)carbamate), and methoxyphenone (3,3′-dimethyl-4-methoxybenzophenone). These herbicides and other chemicals, including antibiotics, were standard commercial products.

Detection of *De Novo* Starch Synthesis

Seeds of barnyard millet and Italian ryegrass were sown in soil in plastic pots and allowed to germinate and grow in a greenhouse at 25 to 30°C. On the 10th day after planting (second leaf stage), the pots were transferred to a dark chamber and kept for 12 hr at 25°C. The second leaf was cut into segments of about 5 mm, and these segments, which were starch negative by iodine staining, were transferred to plastic petri dishes (1 cm diameter, Nunc) containing 1.8 mL of the reaction medium consisting of 0.01 *M* potassium phosphate buffer (pH 6.5) and 0.01% (v/v) Tween-20 (a wetting agent). After adding 0.2 mL of a test solution, the petri dishes were illuminated with an intensity of 14 klux for 16 hr at 25°C. The leaf segments were examined for starch by staining with a solution of 2% iodine and 2% potassium iodide after removal of chlorophyll with hot methanol.

RESULTS

Selection of Plant Leaves Suitable for Starch Assay

A variety of C_3 and C_4 plants were examined for their starch accumulation in leaf tissue in order to select a species suitable for detection of *de novo* starch synthesis. After the plants had been grown under light in a greenhouse, their second or third leaves were examined for starch accumulation by iodine staining. Among C_3 plants, orchard grass (*Dactylis glomerata* L.), Italian ryegrass (*Lolium multiflorum* LAM.), oats (*Avena sativa* L.), radish (*Raphanus sativus* L.), and pea (*Pisum sativum* LINN.) were starch positive by iodine staining, but rice (*Oryza sativa* L. cv. Nihonbare) was negative. Among C_4 plants, common millet (*Panicum miliaceum* L.), foxtail millet (*Setaria italica* BEAUV.), Bermuda grass (*Cynodon dactylon*), barnyard millet (*Panicum crus-galli*), and large crabgrass (*Digitaria ciliaris*) were positive, but maize (*Zea mays* L.) was negative. Italian ryegrass (a C_3 plant) and barnyard millet (a C_4 plant) were used for further starch assay.

Incubation Time Required for Starch Starvation and De Novo Synthesis

The plants were grown in plastic pots to the second leaf stage and then transferred to a dark chamber to remove starch in the leaves, as indicated by negative iodine staining. It took at least 8 hr to remove the starch in every leaf tested. Therefore, all plants were kept in the dark for 12 hr, and then their leaves were cut into segments and allowed to synthesize starch. Photosynthesis was initiated by illuminating the leaf segments, which were floated on the surface of the medium in a petri dish. A period of at least 5 hr was required before synthesized starch could again be detected by iodine staining. The starch synthesis was allowed to continue for 16 hr in order to assay for photosynthesis inhibitors.

Effects of Several Types of Biologically Active Compounds on Starch Synthesis in Leaf Segments of Barnyard Millet

Different kinds of biologically active compounds were examined for their inhibitory action on *de novo* starch synthesis in the excised leaf segments. They included (1) herbicides that inhibit photosynthesis, (2) herbicides that do not inhibit photosynthesis, (3) respiratory inhibitors, (4) protein synthesis inhibitors, (5) RNA synthesis inhibitors, and (6) DNA synthesis inhibitors. Table 1 shows a typical result of the starch assay in leaf segments of barnyard millet. The most potent inhibitors of *de novo* starch synthesis were diuron, propanil, and linuron, which are known as photosynthesis inhibitors. They were detected by this assay at a concentration of 0.1 μg/mL. Although Swep was not as potent as the other three compounds in this group, it was also detected at 1.0 μg/mL. Other active compounds which are not photosynthesis inhibitors included paraquat, which reportedly generates superoxide, sodium azide and 2,4-dinitrophenol, which are respiratory inhibitors, and protein synthesis inhibitors such as cycloheximide, gentamicin, and streptomycin. Among them paraquat, sodium azide, and cycloheximide were detected at 1.0 μg/mL. None of the other compounds, including herbicides, inhibited starch synthesis at a concentration as high as 100 μg/mL.

COMMENTS

Microbial metabolites, in addition to synthetic chemicals, are possible sources of new herbicides. This laboratory has developed an *in vitro* assay method sensitive enough to detect photosynthesis inhibitory activity of microbial culture broths, which may contain very small amounts of active substances. This method was responsive to photosynthesis inhibitors such as diuron, propanil, and linuron at concentrations as low as 0.1 μg/mL. However, other compounds without photosynthesis inhibitory activity also inhibited starch synthesis, especially paraquat, sodium azide, and cycloheximide, which were

Table 1. Effects of Several Types of Biologically Active Compounds on *De Novo* Starch Synthesis in Leaf Segments of Barnyard Millet

	Inhibitory Activity (μg/mL)[a]				
Chemicals[b]	**0**	**0.1**	**1.0**	**10**	**100**
Photosynthesis inhibitors					
Diuron	−	+ +	+ +	+ +	+ +
Propanil	−	+	+ +	+ +	+ +
Linuron	−	+	+ +	+ +	+ +
Swep	−	−	+	+ +	+ +
Nonphotosynthesis inhibitors					
Paraquat (Methylviologen)	−	−	+	+ +	+ +
Nitrofen	−	−	−	−	−
Amitrole	−	−	−	−	−
Chlorpropham	−	−	−	−	−
Glyphosate	−	−	−	−	−
2,4-D	−	−	−	−	−
Methoxyphenone	−	−	−	−	−
Phenazine	−	−	−	−	−
Isoprothiolane	−	−	−	−	−
Sodium azide	−	−	+	+	+ +
2,4-DNP	−	−	−	+	+
Cycloheximide	−	−	+ +	+ +	+ +
Streptomycin	−	−	−	−	+
Gentamicin	−	−	−	−	+
Kasugamycin	−	−	−	−	−
Chloramphenicol	−	−	−	−	−
Antimycin A	−	−	−	−	−
Oligomycin	−	−	−	−	−
Actinomycin C	−	−	−	−	−
Rifamycin	−	−	−	−	−
Bleomycin	−	−	−	−	−

[a] Inhibitory activity was expressed as follows: −, no effect (stained with iodine); +, active (partially stained); + +, active (not stained).

[b] For chemical names of herbicides see references given in the preface of this book.

detected at 1.0 μg/mL. Therefore, it will be necessary to use another method for secondary screening — for example, examining the effect on oxygen evolution using a membrane-coated oxygen electrode or cotyledon disc bioassay — to distinguish between photosynthesis inhibitors and other compounds. Despite this shortcoming, the starch assay presented here seems to be useful for primary screening of photosynthesis inhibitors in microbial culture broths because of its sensitivity and simplicity. Laidlomycin in the culture broth of *Streptomyces* has been found to be a photosynthesis inhibitor using this method. Furthermore, it is expected that this assay method could be used to find an herbicide that shows selection between C_3 and C_4 plants.

REFERENCES

Kida, T., Takano, S., Ishikawa, T. and Shibai, H., A simple bioassay for herbicidal substances of microbial origin by determining *de novo* starch synthesis in leaf segments. *Agric. Biol. Chem.*, 49, 1299, 1985.

Kida, T. and Shibai, H., Inhibition of *de novo* starch synthesis and photosynthesis by laidlomycin. *Agric. Biol. Chem.* 50, 485, 1986.

Kida, T., Screening of microbial products affecting plant metabolism, in *Novel Microbial Products for Medicine and Agriculture,* Demain, A., Somkuti, G., Hunter-Cevera, J. and Rossmoore, H., Eds., Elsevier, Amsterdam, 1989, p. 195.

CHAPTER 37

Screening for Growth Retardants and Herbicides (Conductivity Measurement) with Heterotrophic Plant Cell Suspension Cultures

Klaus Grossmann

INTRODUCTION

Plant cell suspension cultures have found widespread use in recent years in plant physiology and environmental and agrochemical research. Particularly, applications in screening of compounds for bioregulator properties, phytotoxic activity, mutagenic effects, and for activity to elicit tolerance to herbicides (antidotes) have been demonstrated (survey by Grossmann, 1988). However, designing of corresponding test systems presupposes physiological comparability of the target cells in the plant to those in the culture. In this context, heterotrophically cultivated cell suspensions offer an appropriate model system for meristematic tissue, since growth is governed in both cases mainly by cell division activity and both cell species generally contain chloroplasts only as proplastids (Grossmann, 1988). Thus, the usefulness of this sytem has been proved true particularly for determining the effects of plant growth retardants and certain herbicides which are known to primarily influence plant meristems (Grossmann, 1988). For this purpose, a simple and rapid screening method is described which permits an objective monitoring of cell division growth by measuring electrical conductivity in cell suspensions cultured in test tubes.

0-87371-539-X/93/$0.00 + $.50

PROCEDURE

Cell Culture

Freely suspended callus cells from maize (*Zea mays* L. cv. Black Mexican Sweet; kindly provided by Prof. Dr. H. Lörz, University of Hamburg, Hamburg, Germany) and oilseed rape (*Brassica napus* L. cv. Kasan; kindly provided by Dr. R. Hamm, BASF, Limburgerhof, Germany) are cultivated in I2a medium, a modified Murashige-Skoog medium according to Seitz et al. (1985): KNO_3, 1.9 g/L; NH_4NO_3, 1.65 g/L; $CaCl_2 \times 2\ H_2O$, 0.44 g/L; $MgSO_4 \times 7\ H_2O$, 0.37 g/L; KH_2PO_4, 0.17 g/L; Na_2·EDTA, 37.2 mg/L; $FeSO_4 \times 7\ H_2O$, 27.8 mg/L; H_3BO_3, 6.2 mg/L; $MnSO_4 \times 4\ H_2O$, 22.3 mg/L; $ZnSO_4 \times 7\ H_2O$, 8.6 mg/L; KJ, 0.83 mg/L; $Na_2MoO_4 \times 2\ H_2O$, 0.25 mg/L; $CuSO_4 \times 5\ H_2O$, 0.025 mg/L; $CoCl_2 \times 6\ H_2O$, 0.025 mg/L; sucrose, 30 g/L; myo-inositol, 100 mg/L; nicotinic acid, 0.5 mg/L; thiaminium dichloride, 0.1 mg/L; pyridoxine hydrochloride, 0.1 mg/L; L-alanine, 29.7 mg/L; L-arginine monohydrochloride, 3.1 mg/L; 4-aminobutyric acid, 26.0 mg/L; L-asparagine, 3.8 mg/L; L-aspartic acid, 1.7 mg/L; L-cysteine, 3.0 mg/L; L-glutamic acid, 15.7 mg/L; L-glutamine, 0.3 mg/L; glycine, 2.7 mg/L; L-histidine monohydrochloride, 0.05 mg/L; L-leucine, 5.0 mg/L; L-lysine monohydrochloride, 2.0 mg/L; D,L-methionine, 0.05 mg/L; D,L-phenylalanine, 0.05 mg/L; L-proline, 1.9 mg/L; L-serine, 12.8 mg/L; D,L-threonine, 4.1 mg/L; L-tyrosine, 0.05 mg/L; D,L-valine, 2.3 mg/L; 3-indolylacetic acid, 2.0 mg/L; kinetin, 0.2 mg/L; 2,4-dichlorophenoxyacetic acid, 0.1 mg/L; in bidistilled water, adjusted to pH 5.6 with KOH, and autoclaved at 115 °C for 20 min. The cells, grown in culture for several years, are subcultured at intervals of 7 days in the exponential growth phase under the following conditions: 250 mL Erlenmeyer flasks containing 80 mL of cell suspension (with 15 to 20 mL of inoculum) are shaken on a rotary shaker at 110 rpm in the dark at 25°C.

Chemicals

The herbicides and plant growth retardants used (Table 1; for structures see Fedtke, 1982; Grossmann, 1988; Böger and Sandmann, 1990) were obtained from Riedel-de Haën (Seelze, Germany) and BASF Aktiengesellschaft (Ludwigshafen, Germany). Stock solutions of the chemical compounds are prepared 100-fold concentrated (to the final concentration in the test) in acetone.

Test Equipment and Conditions

Freshly subcultured cells from the exponential growth phase (2 mL) are incubated in sterile polypropylene tubes (17 mm diameter, 100 mm height, 15.5 mL volume; Greiner, Frickenhausen, Germany). Before pipetting the cell suspensions using a Pipetman P 5000 with cut tips (C 5000; Gilson,

Madison, Wisconsin) the solutions of test compounds are added to the tubes allowing acetone to volatilize. Final acetone concentrations of up to 1% in the medium have no adverse effects on cell culture growth. The tubes are shaken on a rotary shaker (KS 500, JKA, Staufen, Germany) with an attachment for 288 test tubes, at 400 rpm and 25°C in the dark. After incubation for 8 days the conductivity of the growth medium is measured using a conductivity meter (LF 530, WTW, Weilheim, Germany) with a special microelectrode that is placed directly in the cell suspension in the tube (see photographs in Grossmann, 1988). The mean of three replicates is determined. This value is subtracted from that obtained before incubation. Results are expressed as percent growth inhibition relative to untreated controls. Calculations are performed with a programmed Epson HX-20 attached to the conductivity meter.

In the initial screening of newly synthesized compounds and extracts of natural origin the final concentration applied to the test is 10^{-4} mol · L^{-1} and 60 mg/L, respectively. If growth inhibition exceeds 50%, further tests with decreasing concentrations are performed.

RESULTS

Figure 1 shows that in maize cell suspensions a decrease in conductivity of the medium during incubation is inversely proportional to the increase in cell number. Under the conditions applied, this is independent of culturing the cells in Erlenmeyer flasks or in test tubes. Likewise, the growth cycles in both systems follow a similar pattern. The same behavior has been demonstrated for suspension cultures of sycamore (Grossmann and Jung, 1984). In addition, a comparison of the effects of ten growth retardants on suspension growth in Erlenmeyer flasks measured by cell counting and in test tubes by determining changes in conductivity gave evidence for an accurate recording of the effects by this method (Grossmann and Jung, 1984). Highly significant correlations of the two sytems were obtained in cell suspensions of rice, maize, oilseed rape, sycamore, tomato, and sunflower. When cells were boiled or homogenized at the beginning of the test, released ions increased the conductivity of the growth medium by only 0.25 mS · cm^{-1}. In this case, which mimics the situation after strong damage of the cells by certain herbicides, values of growth inhibition exceeding 100% are obtained.

Comparing the potency of known growth retardants and herbicides using maize cell suspensions, roughly the following order of activity is obtained: N-heterocyclics > cyclohexanetriones > oniums for growth retardants; sulfonylureas > imidazolinones > cyclohexanediones = oxyphenoxypropionic acids > phenoxy acids ≥ herbicides causing radical damage, inhibition of photosynthetic electron flow, and inhibition of mitosis. Interestingly, the responses of cell suspensions of oilseed rape were often less sensitive than those of maize. This is particularly obvious for the two classes of graminicides of cyclohexanedione and aryloxyphenoxypropionic acid type (Table 1).

Table 1. Efficiency of Selected Herbicides and Growth Retardants on Growth of Heterotrophic Cell Suspensions of Maize and Oilseed Rape Determined by Conductivity Measurement after 8 Days of Incubation

Chemical class	Common name[a]	Influence on Plant Process[b]	Molar Concentration Required for a 50% Reduction of Cell Suspension Growth[c]	
			Maize	Oilseed Rape
Herbicides				
Uracil	Lenacil	Photosynthetic electron flow	$>10^{-3}$	$>>10^{-3}$
s-Triazine	Atrazine	Photosynthetic electron flow	5.8×10^{-4}	$>>10^{-3}$
Thiadiazinone-dioxide	Bentazon	Photosynthetic electron flow	2.0×10^{-4}	3.3×10^{-4}
Urea	Chlorotoluron	Photosynthetic electron flow	2.5×10^{-4}	8.5×10^{-5}
Diphenyl ether	Oxyfluorfen	Radical damage	1.1×10^{-4}	1.0×10^{-3}
Bipyridylium	Paraquat	Radical damage	6.5×10^{-6}	1.2×10^{-5}
Carbamate	Asulam	Mitotic process	1.7×10^{-4}	3.6×10^{-4}
Dinitroaniline	Pendimethalin	Mitotic process	5.0×10^{-5}	1.3×10^{-4}
Phenoxy acid	MCPA	Auxin action	3.3×10^{-5}	2.7×10^{-5}
	2,4-D		2.8×10^{-5}	1.7×10^{-5}
Oxyphenoxypropionic acid	Diclofopmethyl	Amino acid biosynthesis	4.8×10^{-6}	1.6×10^{-4}
Imidazolinone	Imazethapyr	Amino acid biosynthesis	4.6×10^{-7}	1.9×10^{-6}
Sulfonylurea	Chlorsulfuron	Amino acid biosynthesis	8.0×10^{-9}	$<10^{-9}$
Cyclohexanedione	Sethoxydim	Lipid synthesis	7.0×10^{-6}	2.7×10^{-4}
	Cycloxydim		4.4×10^{-6}	3.2×10^{-4}
Growth retardants				
Onium	Chlormequat chloride	Gibberellin/sterol biosynthesis	$>10^{-3}$	$>10^{-3}$

Cyclohexanetrione	Prohexadione	Gibberellin/sterol biosynthesis	2.9×10^{-4}	2.3×10^{-4}
Norbonanodiazetine	Tetcyclacis	Gibberellin/sterol biosynthesis	1.2×10^{-5}	4.6×10^{-5}

[a] For chemical names see references given in the preface of this book.
[b] According to Fedtke (1982), Grossmann (1988), Böger and Sandmann (1990).
[c] Applied concentration range 10^{-3}–10^{-9} mol · L^{-1}, four replicated tubes.

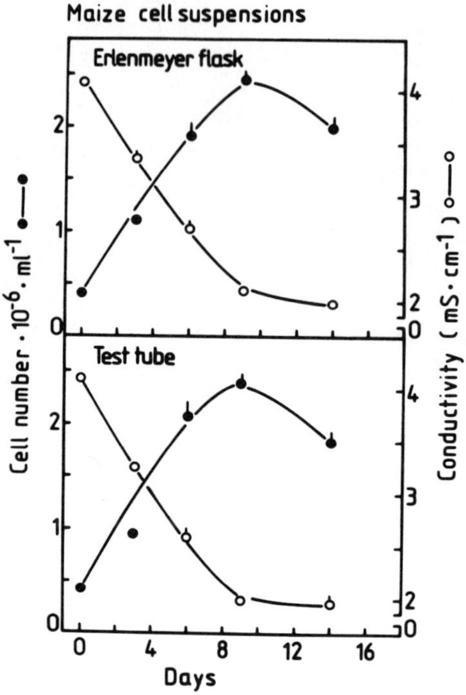

Figure 1. Time course for changes in cell number, determined microscopically and in medium conductivity during incubation of heterotrophic maize cell suspension cultures in Erlenmeyer flasks and test tubes. Vertical bars indicate standard error of the mean (n = 4).

COMMENTS

The determination of the conductivity of the growth medium is a quick, simple, and accurate method of monitoring growth in cell suspensions indirectly. It inversely correlates with increases in cell number, packed cell volume, and cell dry mass (for review see Grossmann, 1988). Thereby, the decrease in medium conductivity during the growth cycle results from an uptake of inorganic ions by the cells. A requirement for this is that the culture medium used contains one or more ionic constituents (e.g., ammonium ions, nitrates, and phosphates), which mainly contribute to the medium conductivity and whose uptake corresponds directly to the growth rate. In combination with the cultivation in test tubes, the method is suitable for cell suspensions of the most species. However, a prerequisite is that the suspensions are homogeneous without forming large clumps which would interfere with pipetting.

By using a series of five table shakers the screening capacity of this micromethod, which has been routinely performed in this laboratory for more than 7 years, is testing 400 compounds applied in one concentration and three replicates a week. Test substances are applied only in microgram amounts, without the need to add detergents and/or other formulation ingredients. As a special advantage, biological activity of compounds with a limited penetration and translocation behavior in the plant can be detected by this test system. This knowledge provides the chance to attempt to solve the problem by modifying the chemical structure or formulation of the compound.

Adapted to the particular function and/or integrated in a battery of screening systems using, for instance, algae, explants, miniplants, and germinating seeds, heterotrophic cell suspensions represent a valuable and even necessary complement to the whole-plant screens routinely used in industrial laboratories. In addition, such biotests are useful in evaluating the physiological action profile of chemicals in search of possible new biological targets.

ACKNOWLEDGMENTS

The author thanks Mrs. D. Meinköhn for technical assistance and Mrs. K. Kolb for typing the manuscript.

REFERENCES

Böger, P. and Sandmann, G., Modern herbicides affecting typical plant processes, in *Chemistry of Plant Protection,* Vol. 6, Bowers, W. S., Ebing, W., Martin, D. and Wegler, R., Eds., Springer-Verlag, Berlin, 1990, p. 173.

Fedtke, C., *Biochemistry and Physiology of Herbicide Action,* Springer-Verlag, Berlin, 1982.

Grossmann, K. and Jung, J., A new micro-method for testing plant growth retardants in cell suspension cultures, *Plant Cell Rep.,* 3, 156, 1984.

Grossmann, K., Plant cell suspensions for screening and studying the mode of action of plant growth retardants, in *Advances in Cell Culture,* Vol. 6, Maramorosch, K. and Sato, G. H., Eds., Academic Press, Orlando, FL, 1988, 89.

Seitz, H. U., Seitz, U. and Alfermann, W., *Pflanzliche Gewebekultur,* G. Fischer, Stuttgart, 1985.

CHAPTER 38

Screening of Peroxidizers that Cause an Accumulation of Photosensitive Tetrapyrroles (Qualitative *Lemna* Assay)

Reynold Chollet

INTRODUCTION

Several chemically unrelated classes of peroxidizing herbicides (e.g., *p*-nitrodiphenyl ethers, *N*-phenylimides, oxadiazon, LS 82556, MB 39279) have the enzyme protoporphyrinogen oxidase (EC 1.3.3.4) as target (see Scalla and Matringe, 1990, for review). The resulting accumulation of a photosensitive tetrapyrrole derivative (eventually protophyrin IX accumulates) induces in the light the formation of free radicals that initiate a peroxidation of lipids followed by membrane disruption and cell death.

A very simple, inexpensive, short, and selective bioassay using *Lemna minor* allows the screening of compounds that cause such an accumulation of photosensitive tetrapyrrole derivatives through inhibition of some step of the porphyrin biosynthesis pathway (see Comment 1).

Principle of the Assay

In the presence of light, peroxidizers cause a photobleaching of the treated fronds. The bleaching is much less pronounced (or even totally absent) either in the dark or in the presence of gabaculine (2-amino-2,3-dihydroxybenzoate),

0-87371-539-X/93/$0.00 + $.50

an inhibitor of the biosynthesis of the porphyrin precursor δ-aminolevulinic acid, (δ-ALA; see Comment 2).

PROCEDURE

Materials

Materials include petri dishes (diameter 20 cm) and microscope lamella boxes (22 × 22 × 18 mm) with perforated lid (one hole, diameter 2 mm; see Comment 3).

Herbicides

Herbicides utilized were acifluorfen-Na: Na-5-(2-chloro-4-(trifluoromethyl)phenoxy)-2-nitrobenzoate; nitrofen: 2,4-dichlorophenyl 4-nitrophenyl ether; S 23121: (1R,S)-(+)-*N*-(4-chloro-2-fluoro-5-((1-methyl-2-propynyl)oxy)phenyl)-3,4,5,6-tetrahydrophtalimide; paraquat: 1,1′-dimethyl-4,4′-bipyridinium dichloride; DNOC: 4,6-dinitro-*o*-cresol; metoxuron: 3-(3-chloro-4-methoxyphenyl)-1,1-dimethylurea; methazole: 2-(3,4-dichlorophenyl)-4-methyl-1,2,4-oxadiazolindine-3,5-dione; norflurazon: 4-chloro-5-methylamino-2-(3-trifluoromethylphenyl)pyridazin-3-one; and metolachlor: 2-chloro-*N*-(2-ethyl-6-methylphenyl)-*N*-(2-methoxy-1-methylethyl)acetamide.

Solvents/Formulations

The following solvents can be used if their final concentration is not higher than 1% (v/v): acetone, isopropanol, xylene, dimethylsulfoxide, or tetrahydrofuran. Ethanol and *N,N*-dimethylformamide should be kept below 0.5%. Methanol and *N*-methyl-2-pyrrolidone should be kept below 0.1%, and acetonitrile should be avoided.

Formulated commercial products can also be used in most cases.

Cultivation

Lemna minor (harvested in a pond and cloned) is grown in mineral medium consisting of KNO_3, 202 mg/L; KH_2PO_4, 50.3 mg/L; K_2HPO_4, 27.8 mg/L; K_2SO_4, 17.4 mg/L; $MgSO_4 \times 7\ H_2O$, 49.6 mg/L; $CaCl_2$, 11.1 mg/L; Na_2-EDTA, 10 mg/L; $FeSO_4 \times 7\ H_2O$, 6 mg/L; H_3BO_3, 5.72 mg/L; $MnCl_2 \times 4\ H_2O$, 2.82 mg/L; $ZnCl_2$, 0.6 mg/L; $Na_2MoO_4 \times 2\ H_2O$, 0.059 mg/L; $CuCl_2 \times 2\ H_2O$, 0.078 mg/L; $CoCl_2 \times 6\ H_2O$, 0.054 mg/L; the pH is adjusted to 6.5.

A nursery culture is maintained in petri dishes (400 mL mineral medium per dish) in a growth chamber at 20°C and under continuous light (6,000 to 8,000 lux). The fronds are washed twice weekly with 0.01 *M* NaClO (maximum 40 to 60 sec; see Comment 4), then rinsed thoroughly with tap water before transferring them into fresh medium. At the same time, larger frond groups are separated into smaller groups (two to three fronds). Only healthy small groups are retained, and the frond population is reduced in order to cover about 10 to 20% of the petri dish surface.

Treatments

Lamella boxes (see section on materials) are filled with 3 mL mineral medium; then healthy pairs of *Lemna* are selected (see Comment 5) and one pair is placed in each box. The boxes are divided into three blocks: **L, D,** and **G.** Each box of the block **G** additionally receives 30 μL of 10 m*M* gabaculine (0.1 m*M* final concentration; see Comment 2). The screening compounds are then pipeted at the desired concentration (see Comment 6) into the boxes of the three blocks, and the perforated lids are placed on top. Block **D** is immediately placed in the dark and blocks **L** and **G** in the light (16-hr photoperiod, 8,000 to 10,000 lux, starting with at least 8 hr light). The growth temperature is 25°C for the three blocks (growth chamber).

Rating

The rating is done visually by comparing the degree of bleaching of the initial frond pairs (see Comment 7) for the three blocks **L, D,** and **G** after 4 days of incubation.

RESULTS

Table 1 shows an example of the protocol for an experiment including several peroxidizing herbicides and some herbicides with other mechanisms of action. Acifluorfen or nitrofen (*p*-nitrodiphenyl ethers) and S 23121 (*N*-phenylimide) are typical peroxidizers that cause an accumulation of the photosensitive tetrapyrrole protoporphyrin IX. Their primary target is the enzyme protoporphyrinogen oxidase (see Comment 1). It is therefore logical that their bleaching activity is strongly reduced either in the presence of gabaculine (no tetrapyrrole biosynthesis) or in the dark (see Comment 8). Paraquat is another type of light-dependent peroxidizer that needs an active photosystem I for causing bleaching (bleaching reduction in the dark), but its activity is independent of the biosynthesis of porphyrins (no bleaching reduction in the presence of gabaculine).

DNOC is known as an inhibitor of both respiration (uncoupling of oxidative phosphorylation) and photosynthesis (photosystem II). At 50 μ*M* bleaching appears only in the presence of gabaculine or in the dark. At 100 μ*M* (result not in table) bleaching appears also in the light. It is then increased in the presence of gabaculine and decreased in the dark.

The light dependence of the activity of the typical photosynthesis inhibitors (e.g., metoxuron, methazole) cannot be detected by this assay because no bleaching occurs during the 4 days of the test even in the light, although growth is strongly inhibited.

Table 1. Protocol for the Detection of an Accumulation of Photosensitive Tetrapyrroles by *Lemna minor* Treated with Different Herbicides

	Bleaching of the Initial Pair of Fronds			
		Strongly Changed[a]		
Herbicide (50 μ*M*)	Present In block L (Light)[b]	In Block D (Dark)	In Block G (+ 0.1 m*M* Gabaculine)	Accumulation of Photosensitive Tetrapyrroles
None	No	No	No	–
Acifluorfen	Yes	Decreased	Decreased	+
Nitrofen	Yes	Decreased	Decreased	+
S 23121	Yes	Decreased	Decreased	+
Paraquat	Yes	Decreased	No	–
DNOC	No (GR)	Increased	Increased	–
Metoxuron	No (GR)	No	No	–
Metazole	No (GR)	No	No	–
Norflurazon	No (WNF)	No	No	–
Metolachlor	No (GR)	No	No	–

[a] When compared to block **L**.
[b] (GR): total growth inhibition without bleaching; (WNF): moderate growth inhibition and whitening of the new fronds.

Also, the carotenoid biosynthesis inhibitors (e.g., norflurazon) cause no bleaching of the initial frond pair. A whitening (= total absence of pigments, not identical with the yellowing caused by gabaculine: see Comment 2) occurs only with new growing fronds (growth is inhibited, but not totally). Metolachlor strongly inhibits the growth of *Lemna minor* without causing any bleaching in all the conditions tested.

COMMENTS

1. Up to now only inhibitors of the enzyme protoporphyrinogen oxidase have been known as herbicides that induce an accumulation of the photosensitive protoporphyrin IX. However, inhibitors of the enzyme protoporphyrin chelatases (especially the chloroplastic Mg-chelatase, the next step in the chlorophyll biosynthesis pathway) should also cause such an accumulation and constitute a new potential class of peroxidizing herbicides. Theoretically the present assay should allow the detection of both types of inhibitors.
2. No new fronds are formed in the dark (block **D**), but the initial pair of fronds remains green and healthy during the 4 days of the experiment. Some new fronds are generally formed in the presence of gabaculine (block **G**), but they remain yellow (no chlorophyll, only carotenoids). Gabaculine can be replaced by 4,6-dioxoheptanoic acid, an inhibitor of δ-ALA dehydrase. However, gabaculine is more effective at a lower rate (0.03 m*M*).
3. Multiple-well plates (e.g., six rows of four holes) can be used in place of single boxes; however, there is the risk of a contamination of several wells by a single highly volatile compound. The lid perforation favors gas exchange and slows down the natural senescence.

4. There is no need for absolute sterile conditions. However, the cultures are often contaminated by undetermined algae that can be maintained at an acceptable level by the NaClO treatment. Higher NaClO concentrations or longer NaClO treatment could be phytotoxic for *Lemna minor*.
5. A *Lemna minor* population includes older (physiologically mature) and younger (immature) fronds. The older and younger fronds may show a different sensitivity to herbicides. A frond pair (two fronds still linked together) always consists of an older and a younger frond; thus, the selection of pairs provides a more homogeneous starting material (if the physiological age and the herbicide sensitivity are considered) than a selection of single fronds, in which case a treatment of two fronds could concern two older or two younger fronds in addition to the previous case (one older + one younger). Further, the separation of pairs from larger groups is much easier than the separation of single fronds (risk of injury).
6. In general, the screening compounds are tested at 50 μM (final concentration). Highly active bleachers are additionally tested at the lowest rate still causing a clear bleaching.
7. The two fronds of the initial frond pair are not always equally bleached because of a possible sensitivity difference between the older and the younger frond (see Comment 5). In that case, only the more damaged frond is considered for the rating.
8. A reduction of the bleaching in the presence of gabaculine is generally linked to a bleaching reduction in the dark. A bleaching reduction in the presence of gabaculine without bleaching reduction in the dark could only be explained by an accumulation of an intermediate of the porphyrin biosynthesis and its light-independent activation (e.g., through the mitochondrial electron transport chain).

REFERENCE

Scalla, R. and Matringe, M., Recent advances in the mode of action of diphenyl ethers and related herbicides. *Z. Naturforsch.*, 45c, 503. 1990.

Indexes

Affiliation Index*

* The numbers indicate chapter numbers. Complete address information can be found in the Contributors section.

General Index